Contents

6 THE ELECTRICAL NETWORK

We have all become so used to having electricity available in our homes at the flick of a switch that it is easy to forget how comparatively recent an innovation it is. Just over a hundred years ago, Swan and Edison invented the incandescent filament lamp — the first practical development of use in the home since Faraday's discoveries of half a century earlier — and in 1881 the first public supply of electricity in the country was established in Godalming, Surrey. In 1890 the country's first power station opened in Deptford, South London, with a 12-mile single-phase cable running to Bond Street and back.

With the introduction of three-phase transmission (still used today), it became possible to transmit electricity over greater distances, and sensible to begin (in 1911) the establishment of an interconnecting system of power stations. However, electricity generation was still a regionalised affair; in 1925 there were 572 separate electricity undertakings drawing their power supplies from 438 generating stations, and

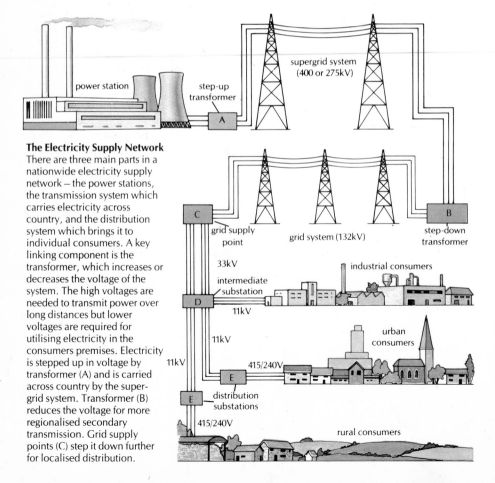

The Electricity Supply Network

There are three main parts in a nationwide electricity supply network – the power stations, the transmission system which carries electricity across country, and the distribution system which brings it to individual consumers. A key linking component is the transformer, which increases or decreases the voltage of the system. The high voltages are needed to transmit power over long distances but lower voltages are required for utilising electricity in the consumers premises. Electricity is stepped up in voltage by transformer (A) and is carried across country by the super-grid system. Transformer (B) reduces the voltage for more regionalised secondary transmission. Grid supply points (C) step it down further for localised distribution.

power station
step-up transformer
A
supergrid system (400 or 275kV)
C
grid supply point
grid system (132kV)
B
step-down transformer
industrial consumers
33kV
intermediate substation
D
11kV
11kV
urban consumers
11kV
415/240V
E
distribution substations
E
415/240V
rural consumers

Electricity

Electricity

Mike Lawrence

Consultant Editor Richard Wiles

OCTOPUS BOOKS

This edition published in 1987 by
Octopus Books Limited
59 Grosvenor Street, London, W1

© 1984 Hennerwood Publications Limited

ISBN 0 7064 2957 5

Printed in Hong Kong

THE ELECTRICAL NETWORK

it took the Electricity Supply Act of 1926 to set up the Central Electricity Board and to establish the National Grid system, which was brought into full commercial operation in 1938.

In 1947 the whole supply industry was nationalised, with a central authority and a number of area boards controlling the supply of electricity to commercial and domestic consumers. The system now consists of around 130 power stations, 9,000 miles of transmission lines and 300,000 miles of distribution lines, and the industry sells the staggering total of over 200,000 million 'units' of electricity a year. When you consider that one unit will run a 100 watt lamp for 10 hours, you get some idea of the enormous scale of the business.

How electricity works

Before you embark on electrical work of any nature, it helps to have a basic idea of how electricity functions. The simplest analogy is with a flow of water through a pipe. Both water and electricity can do 'work' as they flow between two points; water, for example, can turn a water wheel or a turbine, while electricity can produce light if it flows through a lamp, or rotation (which can drive something) if it flows through an electric motor. What causes the flow in each case is a difference in pressure between the two points — the greater the pressure, the greater the flow.

Acting against this pressure is another factor called 'impedance'. Clearly, it's easier for water to flow through a wide pipe than through a narrow one; the latter has the greater impedance to flow, and so the same pressure difference between the ends of the pipe will result in a greater flow through the wider pipe. In electrical terms, impedance (or 'resistance', as it is usually — if incorrectly — called)

divides materials into two types, 'conductors' and 'insulators'. The former have low resistance to the passage of electricity, the latter have high resistance — to the point of not allowing it to pass at all.

Electricity will flow (and carry out work) if it has a circuit to flow round. The flow originates at the source of supply — the power station — and travels through one conductor (the 'live' wire) to wherever it is needed. Then, it returns via another conductor (the 'neutral' wire) to its source. The flow of current (measured in 'amperes' — amps or A for short) is driven through the circuit by the pressure difference (or potential difference, measured in 'volts' — V for short) acting against the resistance of the circuit (measured in 'ohms'). The three are related by the equation: $volts = amps \times ohms$. The amount of electricity consumed whenever work is done (driving a motor or illuminating a lamp, for example) is measured in 'watts' (W), and in practical terms the watts consumed are given by the product of the supply voltage and the current drawn — $watts = volts \times amps$.

In the home, cables deliver electricity to wherever it is needed and then return the flow to the system. Each cable contains a live (flow) conductor, a neutral (return) conductor and a third conductor called the 'protective' or 'earth conductor', or simply 'earth'.

Electricity can escape to earth on its rounds — for example, if you touch a live conductor the current passes through your body to earth, and will shock (or even kill) you. Therefore, everything metallic in your home's wiring system is connected to earth via the earth conductor in the circuit cables. If a fault occurs on an appliance, the voltage on any exposed metalwork cannot rise very high above earth voltage, so will not be dangerous.

8 SAFETY FIRST

The other protection built in to all wiring systems is the 'fuse'. This is a short length of conductor designed to heat up and melt if too high a current passes through it. Such a high current can occur if a circuit is overloaded with too many appliances, if a fault on an appliance occurs that lowers its resistance (such as a short circuit), or if, as already mentioned, electricity is leaking to earth. Fuses are provided in the live pole of all circuits in the house, and in addition are fitted within the rectangular-pin plugs used to link appliances to modern socket outlets. On the most modern systems, the circuit fuses (but not the ones in plugs) may be replaced by devices called 'miniature circuit breakers' (MCBs for short); these isolate a circuit by switching off the current in the event of a fault. In addition, where adequate earthing of the system is difficult to provide, an 'earth-leakage circuit breaker' (ELCB) may also be installed.

Rules and regulations

Unlike most building, plumbing and drainage work (which by law must meet the requirements of the Building Regulations), there are no laws governing how electrical work is undertaken in the home . . . at least, not in England and Wales; in Scotland, electrical specifications *are* covered by the Scottish Building Regulations. However, there are regulations which professional electricians follow — the Regulations for Electrical Installations, published by the Institution of Electrical Engineers, and more commonly known as the IEE Wiring Regulations. It makes sense for the DIY electrician to follow these, too, since not only will the installation be safe but also it will satisfy the Electricity Board when the time comes for your new work to be connected to the mains supply. The Board can refuse to connect you if they believe the condition of your installation could jeopardise the supply to other houses as well as to your own.

It's impossible to summarise the Regulations in a few lines, but some of the most important points, so far as domestic work is concerned, are:
● ring circuits can serve a floor area of up to 100sq m (1076sq ft);
● radial circuits can serve a floor area of up to 50sq m (538sq ft) if wired in 4mm² cable AND protected by a 30A MCB or cartridge fuse, NOT a rewirable fuse; up to 20sq m (215sq ft) if wired in 2.5mm² cable and protected by any 20A fuse or MCB;
● you can connect as many spurs to a ring circuit as there are sockets or fused connection units on the ring itself;
● a spur from a ring circuit can supply only one accessory: a single or double socket, or a fused connection unit;
● in bathrooms, the only socket outlet allowed is a shaver supply unit containing a transformer. Wall heaters and towel rails should be connected to fused connection units but these must not be within reach of the bath or shower cubicle. Lampholders must be fitted with a protective skirt that prevents the metal parts being touched, unless the fitting itself is totally enclosed. Ordinary lampholders may be fitted if they are more than 2.5m (8ft 6in) from the bath or shower cubicle. Wall-mounted switches are allowed only if out of reach of bath or shower; otherwise use cord-operated switches;
● new sockets fitted to power outdoor appliances must have ELCB (earth-leakage circuit breaker) protection — either on the circuit itself, or at the fusebox. The ELCB is designed to detect any slight leakage of current in the circuit being protected, and it will cut off the power supply immediately to prevent accidental electric shock. You'll need expert advice on choosing the right type and model of ELCB.

THE TEN COMMANDMENTS 9

1

Never attempt any electrical work unless you know what you are doing, you understand how to do it and you are confident that you can carry out every stage of the job.

2

ALWAYS turn off the main isolating switch before beginning any electrical work. As an additional safeguard, hang a sign on the switch warning that the supply has been turned off so that no-one will turn it on again.

3

If working on one circuit only, remove the appropriate circuit fuse, or switch off the circuit MCB, before turning on the supply to the other circuits. Keep the fuse in your pocket until you've finished, so no-one can replace it in the fusebox without your knowledge.

4

Double-check all the connections, whether within accessories, plugs or appliances, to make sure that the cores go to the correct terminals, that terminal screws are tight and that no bare conductor is exposed.

5

Never touch any electrical appliance or fitting with wet hands, or use electrical equipment in wet conditions (in particular, out of doors). Never take a portable appliance into a bathroom on an extension lead.

6

Always unplug an appliance before attempting to inspect or repair it.

7

Don't use long trailing flexes or overload sockets with adaptors; fit extra sockets. If you have to extend a flex, use a proper flex extender; don't just twist the cores together and wrap them in insulating tape. If using an extension cable on a drum, always unwind it fully first or it may overheat; check the flex rating if the cable is supplying heaters.

8

Check plug connections and flex condition on all portable appliances at least once a year, remaking connections and replacing flex as necessary. Replace damaged plugs or wiring accessories immediately.

9

Never omit the earth connection. At accessories, take the earth core of the cable to the terminal on the mounting box or accessory as appropriate; on appliances with an earth terminal, use three-core flex and connect its earth core to the terminal.

10

Teach children about the dangers of electricity. The biggest danger areas are socket outlets (fit ones with shuttered sockets to stop them poking in metal objects) and trailing flexes (unplug unused appliances).

10 HOW OLD IS YOUR SYSTEM?

Before you can contemplate making any improvements or alterations to your home's electrical installation, you have got to find out exactly what sort of system exists already. This is particularly important if your house was wired up more than about 30 years ago, because older installations are more difficult to alter and there is also the risk that the system might already be in poor (and potentially dangerous) condition. This is because the materials used to insulate the cables deteriorate with age; once the insulation has failed, the whole system poses safety and fire risks.

Identifying cables

You can easily estimate the age of your system by looking for certain tell-tale signs. The most important, as already mentioned, is the cable itself. The earliest wiring systems used separate stranded copper wires, which were run round the house in surface-mounted steel conduit, lengths of which were screwed together and joined to cast-iron boxes containing light switches and, later, socket outlets for electrical appliances. Somewhat later, lead-sheathed cable containing two rubber-covered stranded copper wires was introduced, with the metal sheath providing earth continuity round the system. Steel boxes housed switches and socket outlets. Then, tough rubber-sheathed (TRS) cable, containing two insulated wires (called cores) and a bare earthing core, took over as the most common type, and plastic switches and sockets began to replace metal ones as plastic moulding techniques improved. Finally, PVC-sheathed cables, used on all modern wiring installations, came along in the late 1940s and early 1950s.

Since you are probably not the first person to want to extend your house installation, you may find there's a mixture of cable types in different parts of the house. The way to find out if this is the case is to turn off the power to the whole house at the main switch and to open up several light switches and socket outlets so that you can examine the cables running into them. Remember that conduit systems may have had new cables run inside them to replace the originals. It is also a good idea to check the condition of the cable insulation at the same time; if it is showing signs of deterioration, you should consider rewiring the whole system.

The second sign of your system's age will be found at the main fuseboard where the electricity supply enters the house — usually near the front door, but occasionally in a cupboard in a front room alcove. In an older installation, the cables from the electricity meter will pass to a main metal distribution box containing an on-off switch for the whole system and also several pairs of fuses in porcelain fuseholders. From this box, more cables feed similar but smaller sub-distribution boxes; two (or more) of these will supply the socket outlets in the main rooms, while at least one will feed the house lighting. A separate box, linked directly to the meter, may be fitted to supply the much heavier current needed for an electric cooker. You will notice that, throughout the system, a fuse is placed in both the live and neutral cables or 'poles' — so-called double-pole fusing.

Fuses and accessories

In a more modern installation, all these distribution boxes will have been replaced by a single box called a 'consumer unit'. This is a metal or plastic enclosure that contains the main system on-off switch and a number of single fuses. Each of these fuses protects one circuit, and is placed in the live pole only; all the neutral cores of the circuit cables are linked to the unit's main neutral terminal, the earth cores to the main earth terminal.

The third (and least reliable) sign is the type of light switches and socket outlets fitted. The earliest switches you're likely to find will be round brass or brown Bakelite toggle types, probably mounted on small wooden blocks called 'pattresses'. Early socket outlets will probably be brown or cream plastic, again mounted on wooden blocks, and will have round holes. There were three sizes of socket outlet rated at 2, 5, and 15A; the smallest was for lamps and radios, the second for electrical appliances such as vacuum cleaners, and the biggest for heavy current-users such as electric fires.

Each type of socket was wired with cable of the appropriate size, a move intended to save running expensive heavy-duty cables to outlets feeding only light-duty equipment, and different-sized plugs were needed for each socket — they were not interchangeable. Since the 1950s, sockets rated at 13A with three rectangular holes have been used instead (although round-hole sockets are still available, and the smallest one is a useful way of providing power solely for table and standard lamps in a modern installation). There

is only one size, designed to take the familiar flat-pin plug with its own internal fuse (round-pin plugs were not fused). If you have this type of socket, don't automatically assume that your whole system is modern; the old sockets may have been replaced by new ones, but the cables connected to them may still be original.

When inspecting an old wiring system, look out for tell-tale signs that indicate the need for a re-wire. These include double-pole fusing (above), switches on wooden pattresses and old roses and socket outlets (below).

12 MODERN ELECTRICAL CIRCUITS

The basic electrical circuit, as we have already seen, consists of two wires. The current flows along one to wherever it is needed, and back along the other to its source. In house wiring circuits, the basic principle is the same, with electricity passing along the 'live' core of the cable and back down the 'neutral' core; the third core in modern cable is used to provide earthing continuity throughout the system.

Each circuit starts at the consumer unit or distribution box, where the live core of the cable is connected to the live terminal of one of the fuseways, and terminates at the neutral terminal. Most circuits are wired as 'branch' lines, with the cable feeding one or more outlets along the branch; whether this outlet is a light fitting, a socket outlet or a single major appliance, the live core loops in and out of one terminal of the outlet while the neutral core loops in and out of the neutral terminal, until the last branch outlet.

Ring circuits

The one exception to this is the 'ring' circuit, used on modern installations to provide power to socket outlets. This is a 'loop' line rather than a branch line, with a continuous cable running out from the consumer unit and returning to it. The two ends of the live core are both connected to the live terminal of the fuseway protecting the circuit; this means that electricity can flow to a socket outlet on the ring along two routes instead of one, and in practice allows the circuit to provide power to more sockets than a single branch line could. In effect, one modern ring circuit takes the place of a number of branch (or 'radial') circuits, with obvious savings on the cost of cable. However, radial power circuits are still used to take power to single major appliances such as cookers (page 56), instantaneous showers (page 58) and immersion heaters (page 60), and also to outbuildings (page 62). Another use is to feed socket outlets in remote parts of the house where running a ring would mean a waste of cable.

Lighting circuits

Most houses have at least two lighting circuits, one for upstairs and one for downstairs. These are wired as radial circuits. What complicates them is the need to provide switching for each light at a convenient point in each room, and there are two ways of doing this so that the switch can 'break' the live core of the circuit cable and, in doing so, turn off the light.

One way is to fit a four-terminal 'junction box' (or joint box, as it is sometimes called) at an appropriate point on the circuit cable; this allows the supply to continue on to the next lighting point while providing connections for one cable to run down to the switch position and another to run to the light controlled by that switch. The light cable is then connected to the terminals of a 'ceiling rose', and the lampholder is connected to the rose by a piece of flexible cord (flex — see page 18). This is called the junction-box system; as you can see from the diagram, it could use up a lot of cable in certain situations.

The second and more recent way of wiring up lighting circuits uses ceiling roses containing extra terminals. The circuit cable simply runs from rose to rose instead of from junction box to junction box, and terminates at the furthest rose on the circuit. Only one

Right: Lighting circuits are wired up as branch lines or radial circuits, with a cable running from a 5A fuseway in the consumer unit to a number of connection points or outlets along the circuit. These may be junction boxes (A), from each of which run separate cables to a light fitting and to its associated switch, or loop-in ceiling roses (B) which act as junction boxes as well.

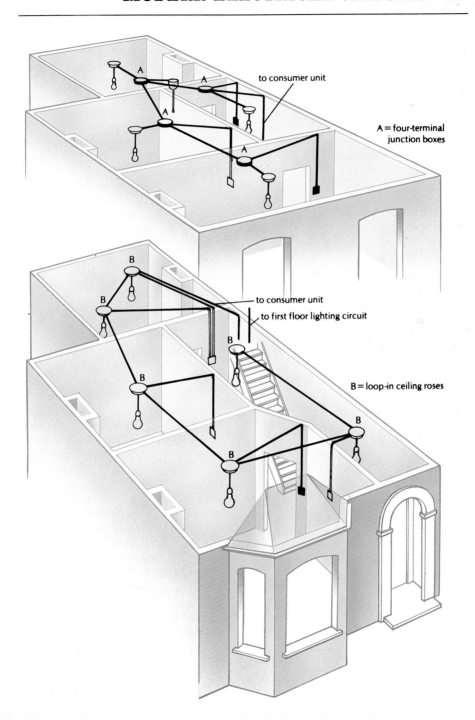

to consumer unit

A = four-terminal junction boxes

to consumer unit

to first floor lighting circuit

B = loop-in ceiling roses

14 MODERN ELECTRICAL CIRCUITS

extra cable is needed, to link the rose with the switch controlling it. This is the loop-in system, so-called because the cable loops in to one rose and out to the next.

Many houses are wired up with a mixture of both loop-in and junction box wiring, because maximum economy of cable can be achieved in this way. It all depends on the positions of the light and switch in each room as to which system is the best to choose.

In a modern installation, each lighting circuit is protected by a 5A fuse or MCB in the consumer unit, and is wired up in 1 or 1.5mm^2 cable — see page 18. Therefore, each circuit can supply up to 1,200 watts — say, twelve 100W light bulbs. In practice, because you may want to fit some bulbs of higher wattage, each circuit is restricted to eight lighting points, so in a large house several lighting circuits may be needed.

Power circuits

As explained previously, modern power circuits to socket outlets are almost always wired up as ring circuits. The cable loops in and out of the back of each socket in turn, eventually returning to the consumer unit. One ring circuit can have as many socket outlets on it as you like; since it can carry current in two directions, and you are unlikely to want to use all your appliances at once, it's unlikely to be overloaded. However, kitchen outlets might warrant a separate circuit. The only restriction placed on the installation is that one ring circuit must not serve a floor area greater than 100sq m (1076sq ft). What's more, you are allowed to increase the number of socket outlets served by adding branch lines or spurs, each being a single cable connected to the ring at an existing ring socket outlet (or at a junction box cut into the ring cable) and taking power to

one single socket, one double socket or one fused connection unit. You are allowed as many spurs as there are sockets or connection units on the original ring. Before 1983, *two* singles or one double socket were permitted on each spur, but new regulations have outlawed the second single socket.

Each ring circuit is protected by a 30A fuse or MCB, and is wired in 2.5mm^2 cable. Therefore, each can supply up to 7,200 watts. In addition to the protection provided by the circuit fuse, each plug or connection unit used to link appliances to the circuit contains a fuse which will 'blow' in the event of a fault on the appliance.

As mentioned earlier, you can also supply socket outlets via radial circuits in situations where a ring circuit would be uneconomical in use of cable. Again, there are no restrictions on the number of socket outlets on each circuit (new regulations changed the old rules), but there are restrictions on the floor area served. For a floor area not exceeding 20sq m (215sq ft), the circuit must be run in 2.5mm^2 cable and be protected by a 20A circuit fuse. For a floor area of up to 50sq m (538sq ft), a 30A MCB or cartridge fuse and 4mm^2 cable must be used.

Other power circuits to fixed appliances are quite straightforward — suitably-sized cable running from a fuseway in the consumer unit (where a fuse or MCB of the appropriate rating is fitted) to the appliance concerned. See pages 56 to 65 for more details.

Right: In modern installations, power circuits are wired up as a ring, starting and ending at a 30A fuseway in the consumer unit and looping into and out of socket outlets on the ring circuit. Branch lines or spurs can be connected to the ring circuit either at a ring circuit socket (A) or at a 30A junction box (B) – see text. Some power circuits – for example, to higher-rated appliances such as cookers, showers and immersion heaters – are wired up as radial circuits.

MODERN ELECTRICAL CIRCUITS 15

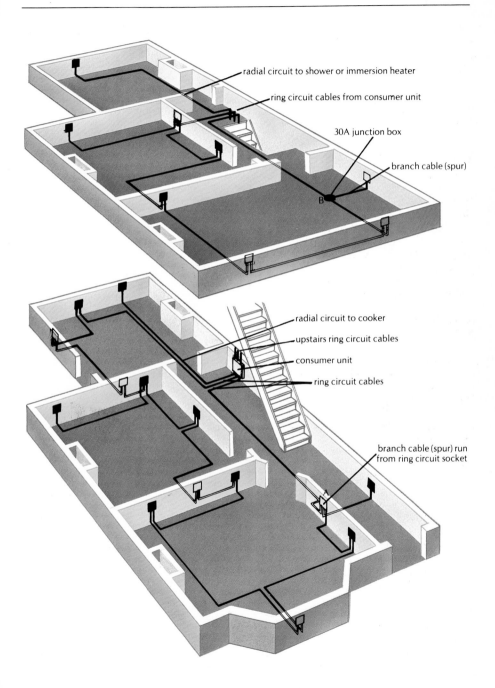

radial circuit to shower or immersion heater

ring circuit cables from consumer unit

30A junction box

branch cable (spur)

radial circuit to cooker

upstairs ring circuit cables

consumer unit

ring circuit cables

branch cable (spur) run from ring circuit socket

16 LIGHT AND POWER FITTINGS

On lighting circuits, most lights take the form of pendant lampholders linked by flex to surface-mounted ceiling roses. In certain situations, the light bulb is fitted into a batten lampholder that is fixed directly to the ceiling or wall. Switches for turning these lights on and off come in a variety of types and sizes. The commonest type may have one, two, three, four or even six switches on a single faceplate. Each switch or 'gang' may have two or three terminals on the back; the former type is used for one-way switching, the latter for two-way switching (see pages 36 and 37). Slimmer versions of these fittings, called architrave switches, are intended for fitting on door frames.

Specialist switches include dimmers (which vary the light level from fully on down to about 10 per cent) and time-delay switches (which turn a light off automatically at a preset time after it has been switched on). There are also cord-operated ceiling switches and weatherproof switches for use outside (see pages 62 to 65).

On power circuits, the most widely used accessory is the socket outlet, in single (one-gang), double (two-gang) and occasionally triple (three-gang) versions. The outlet may be switched

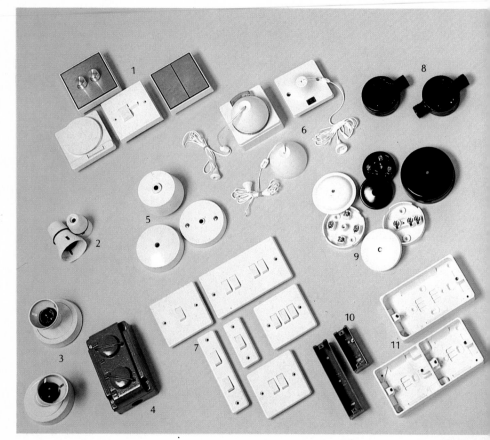

or unswitched, and may include a neon light to indicate when the socket is on. Most modern installations are fitted exclusively with sockets to take flat-pin plugs, but sockets for round-pin plugs are still available, and the smallest (2A) version can be useful on a fused spur circuit feeding table or standard lamps only. Fused connection units are used to connect up non-portable appliances such as freezers, electric towel rails and waste disposal units that are not unplugged regularly (and, indeed, should not be unplugged accidentally). Double-pole switches are used to isolate fixed appliances such as cookers, immersion heaters and instantaneous showers (an ordinary switch is single-pole, interrupting only the live pole). There are also special connection units for cookers, shavers, clocks and television aerials.

1: dimmers; 2: lampholders; 3: batten lampholders; 4: outdoor switch; 5: roses; 6: ceiling switches; 7: plateswitches; 8: BESA boxes; 9: junction boxes; 10: flush boxes for architrave switches; 11: surface-mounting boxes; 12: TV and phone outlets; 13: plug-in timer; 14: connection units; 15: cooker units; 16: blanking plate; 17: cable extender; 18: shaver adaptor and 3-pin plugs; 19: socket outlets; 20: shaver supply unit; 21: flush boxes.

18 CABLE, FLEX AND LAMPS

It is important not to confuse cable and flex. Cable is used for all the fixed wiring on every circuit, running between the consumer unit and the various accessories fitted round the circuits. Flex is used to link appliances of all types to the fixed wiring — via plugs at socket outlets, between ceiling roses and lampholders at lighting points.

Cable for house wiring usually contains three cores. Two are insulated in colour-coded PVC — red for the live core, black for the neutral; the third core is a bare copper wire used to form the earth continuity conductor round the circuit. All the cores are covered with a thick PVC outer sheath; such cable is properly referred to as 'two-core and earth PVC-sheathed and insulated cable', and is usually grey or white. There is also special three-core and earth cable, used in two-way switching installations. Common cable sizes, measured by the cross-sectional area of the copper conductors, are 1, 1.5, 2.5, 4 and 6mm^2.

Flex usually contains three cores as well, all with colour-coded insulation — brown for live, blue for neutral and green-and-yellow for earth. You may find these cores coded red, black and green respectively in old flex. The outer sheathing of flex is normally white or coloured PVC, although a braided cover may be added. Two-core flex has no earth core, being used for lights and pendants with no metallic parts and double-insulated appliances such as power drills, garden tools, hair dryers and food mixers. Special heat-resisting flex is used in powerful light fittings and to connect equipment such as immersion heaters.

Colour blindness may hinder the identification of the appropriate cable and flex cores when making connections. If you have difficulty in distinguishing colours, match the appearance of the wires with those shown in the wiring diagrams in this book.

Light bulbs and fluorescent tubes

Most light fittings take ordinary tungsten-filament light bulbs — properly called GLS (General Lighting Service) lamps. These have a two-pin 'bayonet'

From left: five sizes of two-core and earth PVC-sheathed and insulated cable (6, 4, 2.5, 1.5 and 1mm^2); three-core and earth cable; circular three-core flex; braided three-core flex; circular two-core flex; flat two-core flex; parallel twin flex. Note the uninsulated cable earth cores.

CABLE, FLEX AND LAMPS 19

cap, may be of clear or obscure (pearl) glass — this can also be coloured — and come in two common shapes; round and mushroom. There are also small candle, ball, pygmy and strip lamps for use in wall lights and other decorative fittings. For spot and flood lights, there are special internally-silvered (IS), crown-silvered (CS) and parabolic aluminised reflector (PAR) lamps, of plain and coloured glass. Many of these small and specialised lamps have a threaded Edison Screw (ES) cap instead of the more familiar bayonet fitting, and need special screwed fittings to hold them. A wide range of wattages is available.

Fluorescent tubes are usually 25 or 38mm (1 or 1½in) in diameter, and up to 2.4m (8ft) long, although narrower and shorter 'miniature' tubes are also available. You can even get circular fittings. Wattage depends on tube length — roughly 10 watts per 300mm (12in). The light emitted is described in terms such as 'daylight', 'white' or 'warm white'; coloured tubes are made as well. Most tubes have a bi-pin cap at each end, and are usually mounted in fittings that incorporate a starter (to initiate the discharge through the gas in the tube) and other components to control the discharge once the tube is lit. There may be a diffuser over the tube, or reflectors along both sides.

1: decor round lamps; 2: candle lamps; 3: pearl lamps with bayonet fittings; 4: internally-silvered (IS) lamps; 5: crown-silvered (CS) lamps; 6: double-cap filament and fluorescent miniature tubes; 7: fluorescent tubes; 8: circular fluorescent tubes; 9: coloured lamps; 10: PAR lamps; 11: SL lamp.

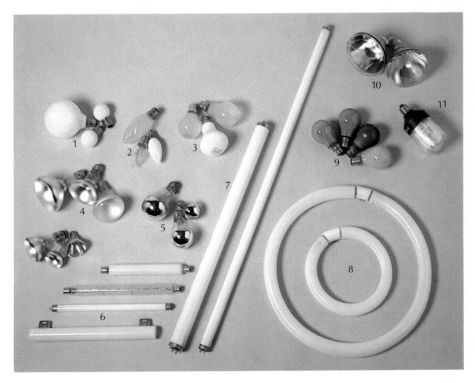

20 TOOLS FOR ELECTRICAL WORK

Even if you never intend to carry out any electrical installation work in your home, you should still have a small tool kit to enable you to cope with electrical emergencies and occasional jobs such as rewiring a plug or fitting a new length of flex to a light or other appliance.

This tool kit should be kept together in a box or bag, ideally next to your consumer unit so you can find it when you need it. You should include: a torch; fuse wire or cartridge circuit fuses to match the ratings of your light and power circuits; 3A and 13A fuses for flat-pin plugs; a roll of PVC insulating tape; a small electrical screwdriver with a sheathed blade; a sharp trim-ming knife; and a pair of wire strippers.

If you are likely to carry out some of your own wiring, you'll need: a pair of pliers (they will cut cable, too, if you pick a pair with cutting jaws; otherwise, add a pair of side cutters as well. Both should have insulated handles); a mains tester for checking that your connections are correct; brick bolster and cold chisel, plus a club hammer, for chopping out cable runs and reces-ses for flush-mounted accessories; power drill (with extension cable; the power may be off in the area where you are working) or brace and bits, for drilling through joists, and hand drill for drilling fixing holes in walls; a floorboard saw or circular saw for

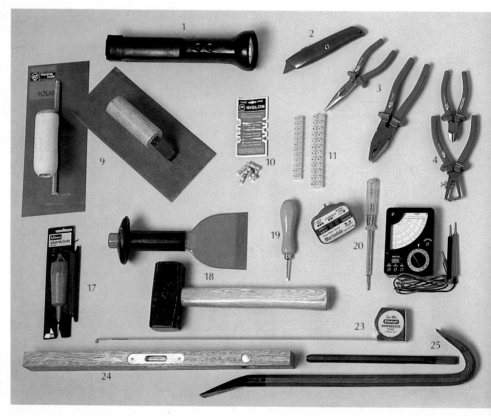

TOOLS FOR ELECTRICAL WORK 21

cutting through floorboards. In addition, you will need a number of everyday tools — for example, screwdrivers, chisels, hammers and saws, all of which you probably own already.

You will also need some cable clips (to hold cable in channels, or chases, cut in walls before you plaster over them, and to support cable runs along the sides of joists), and some green/yellow PVC sleeving (whenever you expose the cable's bare earth core at an accessory, you must cover the bare conductor with this sleeving). Some metal or plastic channelling to protect cables buried in plaster will be found useful, but it is not an essential part of the kit.

1: torch; 2: trimming knife; 3: two sizes of pliers with insulated handles; 4: side-cutters and wire strippers; 5: screwdrivers; 6: wood chisels; 7: assorted twist drills, flat bits and masonry drills; 8: electric drill; 9: plastering floats; 10: fuse wire and plug cartridge fuses; 11: strips of terminal connectors; 12: wood screws; 13: wall plugs; 14: green and yellow striped earth sleeving and red PVC insulating tape (needed for 'flagging' switch returns); 15: mini hacksaw; 16: tenon saw; 17: plumb bob and line; 18: brick bolster (with hand guard) and club hammer; 19: bradawl; 20: ring circuit tester, neon screwdriver and test meter; 21: claw hammer; 22: padsaw; 23: retractable steel tape measure; 24: spirit level; 25: cold chisel and crow-bar; 26: cable clips; 27: PVC grommets (for use with flush mounting sockets to prevent cables chafing where they enter the box); 28: extension cable on drum; 29: floorboard saw.

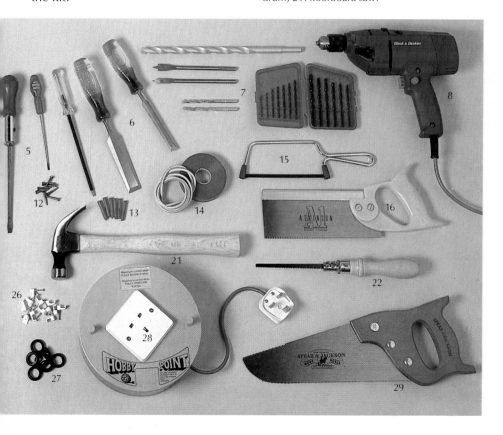

22 DESIGN AND PLANNING

Before you begin to tackle any electrical work, it pays to stop and think about two points — design and ergonomics.

To most people, even a modern wiring accessory is just an uninspiring blob of white plastic that is functional rather than decorative. However, there are ranges of more attractive fittings around in coloured plastic, brushed aluminium, ornate brass and so on, all worth considering if you are prepared to pay more for your accessories. Lightswitches are particularly noticeable because they are less likely to be concealed, and money spent on updating these will certainly be well spent. You can cheer up portable appliances, too, by fitting them with coloured curly flex and matching coloured plugs.

Perhaps more important than the looks of your accessories is the question of ergonomics — making the system work in harmony with you. You must first decide what demands you're likely to place on your system, which appliances need a socket or other outlet full time and which ones can 'borrow' a socket when they're needed. You should also think about positioning light switches, pendant and wall lights and socket outlets for maximum convenience.

First, light switches. These are usually fitted beside the main door to each room, about 1.4m (4ft 6in) above the floor. Fine; but would it be worth having a second switch elsewhere, such as by the patio doors leading to the garden, or by the back door in the kitchen? What about wall lights; should they be switched by the door, by the light or at both points? What about two-way switching for the landing, stairs and hall, for the bedside lights, or even for the immersion heater (a remote switch in the kitchen could save unnecessary trips upstairs)?

Light positions need careful thought, too (although lighting design is a highly specialised field beyond the scope of this book). The central pendant fitted in most rooms is only any good for general lighting; it's hopeless for reading or sewing, or for illuminating dark corners. Would another position be more useful? Would a dimmer control help? What about a rise-and-fall fitting over a dining table?

So far as socket outlets are concerned, you can never (within reason) have too many, and it costs little to add more, either to an existing system or as part of a rewiring job. Remember that there are no restrictions on the number of sockets you can have on a ring circuit. As a guideline to numbers, ignore the Parker-Morris standards still followed by so many architects and specifiers; they were drafted 30 years ago and are pathetically out of date. Instead, aim to match the recommendations of the EIILC (Electrical Installation Industry Liaison Committee), which suggests the following numbers of *double* switched socket outlets:

Room	No. of socket outlets
Living room	6
Dining room	3
Kitchen	4
Hall	1
Landing/stairs	1
Double bedroom	4
Single bedroom	3
Bedsitting room	4
Garage	2

Remember, too, that they don't all have to be set at skirting board level. Some may be more useful at about 750mm (2ft 6in) above the floor — for appliances in alcoves or near fitted units in living rooms, and for freestanding appliances in kitchens — and sockets at this height are far more accessible. In kitchens, some sockets should be above worktop level.

DESIGN AND PLANNING 23

Supplying power to domestic appliances

Appliance	Needs	See pages
Kitchen		
Washing machine	Connection to ring main or spur; ideally, switched fused connection unit (13A fuse)	24, 30, 48, 50
Instantaneous water heater, chest freezer, dishwasher, waste disposal unit	Switched fused connection unit (13A fuse) on ring main or spur	24, 28, 30, 48, 50
Extractor fan	Switched fused connection unit (3A fuse) on ring main or spur; clock connector to link appliance flex to cable	24, 28, 30, 48, 50
Cooker	Separate 30A radial circuit from consumer unit to cooker control unit; cooker connection unit to link appliance trailing cable to supply cable	24, 28, 52, 56
Refrigerator, kettle, toaster, food mixer and other portable appliances	13A switched socket outlet	24, 28, 30, 46, 48
Bathroom		
Instantaneous shower	Separate 30A radial circuit from consumer unit; 30A double-pole pull-cord switch for appliance control (up to 7kW)	26, 28, 52, 58
Extractor fan	As kitchen above	
Immersion heater	Separate 15A or 20A radial circuit from consumer unit; control by 20A double-pole switch with flex outlet and neon indicator	24, 28, 30, 48, 50, 52, 60
Heated towel rail	Switched fused connection unit (13A fuse) on ring main or spur; flex outlet to link appliance to cable in small bathrooms	24, 28, 30, 48, 50
Wall heater	Switched fused connection unit on spur from 30A three-terminal junction box or socket outlet on ring main (also flex outlet in small bathrooms)	24, 28, 30, 48, 50
Living/dining rooms		
Wall lights	Junction box on lighting circuit or switched fused connection unit (5A fuse) on ring main or spur	28, 32, 40
Standard/table lamps	Socket outlets on ring main or spur; 2A round-pin versions suitable (spur must be fused)	24, 28, 30, 46, 48
TV aerial	Aerial socket sited near TV; coaxial cable link from socket to roof-mounted aerial	24, 28, 67
TV, stereo and other portable appliances	13A switched socket outlets	24, 28, 30, 46, 48
Miscellaneous		
Loft light	Spur from four-terminal junction box on lighting circuit; one-gang, one-way switch on landing	26, 28, 32, 34
Central heating controls	Socket outlet or switched fused connection unit (5A fuse) on ring main or spur; ideally, separate 5A radial circuit from consumer unit	24, 28, 30, 48, 50, 52
Door bells	Via transformer fed by separate 5A radial circuit from consumer unit or spur from lighting circuit	28, 32, 52, 66
Outside lights	Spur from four-terminal junction box on lighting circuit	28, 32, 34
Garage	Separate radial circuit from consumer unit; separate switchfuse unit to control power and lighting circuits in outbuilding	24, 26, 28, 34, 48, 50, 52, 62, 64
Garden	For temporary use: extension flex plugged into ELCB-protected socket outlet. For permanent use: separate radial circuit from consumer unit or switchfuse unit, or low-voltage supply via plug-in transformer for pond pumps, etc	30, 50, 52, 64

24 FIXING WALL-MOUNTED ACCESSORIES

You can either mount your switches, socket outlets and other accessories on the wall surface, or recess them into the wall. The former is quicker and easier, but the latter looks much neater and is worth the extra effort (except, for example, in outbuildings where appearances matter less).

Boxes for surface-mounting are usually plastic or metal (to match heavy-duty 'metalclad' accessories) and are simply screwed to the wall where needed (below), using screws driven into wallplugs (solid walls) or cavity fixings (partition walls; alternatively, the box can be screwed to a vertical timber stud or horizontal noggin, if one is conveniently placed). Boxes for flush-mounting are of galvanised steel, and are screwed to the back of the recess in which they fit, as shown opposite. Any gaps between box and masonry are then made good with plaster or filler.

Most light switches are designed to fit a box only 16mm (⅝in) deep. The flush-mounted type is called a plaster-depth box, because in most houses only the plaster has to be cut out to fit them — far simpler than tackling solid brickwork.

Socket outlets, whether surface- or flush-mounted, usually need boxes 35mm (1⅜in) deep, but shallower boxes only 25mm (1in) deep coupled with accessories with thicker-than-usual faceplates are used in single-brick and cavity walls where chopping a deeper recess could result in a hole right through the wall. Certain special accessories such as cooker control units and shaver units need deeper mounting boxes — often 45mm (1¾in) in depth.

Boxes of both types have 'knockouts'; weak areas that are literally knocked out as necessary to allow cables to enter the box from any direction. With flush-mounted boxes, the cables usually enter through the

1 After feeding in the supply cable, remove one of the knockouts from the base of the mounting box. Then hold the box in place and mark the screw positions on the wall.
2 Drill and plug the wall, and then screw the box in place with two woodscrews. Check that it is level.
3 Sleeve the bare earth core with green/yellow PVC sleeving and then connect the live, neutral and earth cores to the terminals on the back of the accessory's faceplate. Make sure the insulation reaches right up to the terminals, and tighten the terminal screws fully.
4 If possible, push excess cable back into the wall or floor cavity, or fold it carefully into the mounting box. Push the faceplate into position and attach it to the box with the fixing screws provided.

1
2
3
4

FIXING WALL-MOUNTED ACCESSORIES 25

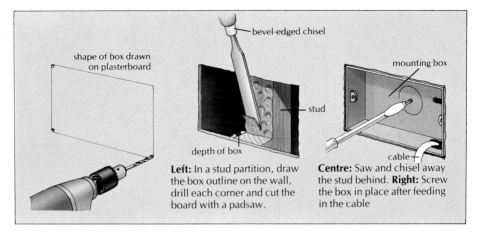

shape of box drawn on plasterboard

bevel-edged chisel

mounting box

stud

depth of box

cable

Left: In a stud partition, draw the box outline on the wall, drill each corner and cut the board with a padsaw.

Centre: Saw and chisel away the stud behind. **Right:** Screw the box in place after feeding in the cable

sides of the box (and must be protected from chafing on the metal edges by a rubber grommet fitted when the knock-out is removed).

With the surface-mounted boxes, the cable will enter through the back of the box if it is run below the wall surface, through its sides otherwise —

see pages 28 and 29 for further details.

You need a one-gang (square) mounting box for single socket outlets, connection units, 20A switches, and one- two- or three-gang plateswitches; a two-gang (rectangular) box for double sockets and for four- or six-gang plateswitches.

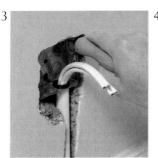

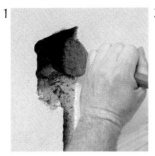

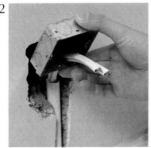

1 With a masonry wall, use a club hammer and brick bolster to cut out a neat chase for the cable, and to chop out a recess to match the size of the flush box being used. Drilling round the perimeter of the hole first with a masonry drill may ease the job on hard brickwork.
2 Lay in the cables, ideally in PVC conduit. Then remove a knockout from the side of the box, insert a grommet to stop the box chafing the cable and feed in the cable.
3 Check that the box fits snugly and squarely within the recess. Chop out more masonry if it's too shallow, fill with a dab of plaster if it's too deep. Then mark the wall for the mounting screws.
4 Remove the box, drill and plug the holes, replace the box and screw it firmly to the wall. Finally, make good the wall.

26 FIXING CEILING-MOUNTED ACCESSORIES

Fixing accessories to ceilings poses slightly different problems to mounting fittings to walls. For a start, such accessories are usually surface-mounted. Most ceilings are 'hollow' — like partition walls, with lath-and-plaster or plasterboard forming the ceiling surface — but the types of fixings used there can't be relied upon to support the weight of heavy light fittings. So any ceiling rose or fitting must be screwed to a proper support. This can be either a joist, or else a batten fixed between two joists. In the former case, the joist might not coincide with your required location. In the latter case, you will have to lift floorboards in the room above (or go up into the loft) so you can fix the batten.

Light fittings, other than simple pendants, have open-back ceiling plates, and the flex from the light is usually linked to the circuit cable via a multi-way cable connector, which can cope with loop-in wiring. This connector has

to be housed within a non-combustible enclosure, consisting of a circular metal or plastic box (called a BESA box, or terminal conduit box) and the fitting's own ceiling plate, which is screwed to the ceiling or to the lugs of the BESA box itself (with M4 metric machine screws).

This BESA box is fitted flush with the ceiling surface, and is usually screwed to the underside of a batten fixed between the joists; in some cases, it may be possible to cut away the underside of the joist slightly and screw the box directly to it, but this is not recommended unless the joist is generously thick. In general, the side-entry type of BESA box is the most useful type to use.

Batten lampholders are either fitted on a backplate (called a pattress) which allows space for the connections, or else over a BESA box, as above. Modern loop-in batten lampholders have an enclosed backplate, so a BESA box is not needed.

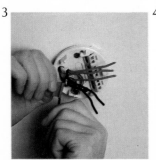

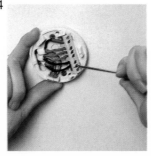

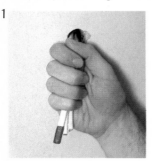

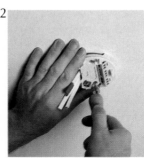

1 Draw the cable (or cables) down through a hole in the ceiling where the rose is to be fitted. Here, the rose will be wired as an intermediate rose on a loop-in system, so there are two circuit cables and one switch cable, identified with red PVC tape.
2 Remove a knockout from the base of the rose, feed the cables through and mark the positions of the fixing screws.
3 Wrap some red PVC tape round the switch cable neutral to show that it is live when connected up.
4 Connect the three live cores to the centre group of terminals, the sleeved earth cores to the earth terminal and the two neutral cores from the circuit cables to the neutral terminal bank. Link the switch cable neutral to the live load terminal at the end of the bank.

FIXING CEILING-MOUNTED ACCESSORIES 27

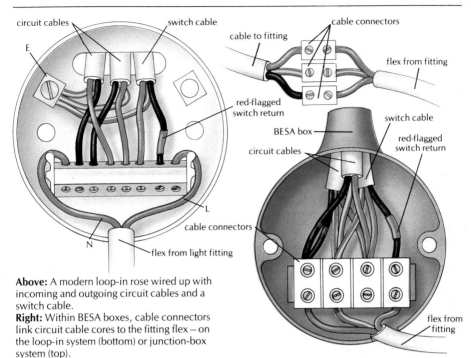

circuit cables

switch cable

E

cable to fitting

cable connectors

flex from fitting

red-flagged switch return

switch cable

BESA box

circuit cables

red-flagged switch return

L

cable connectors

N

flex from light fitting

flex from fitting

Above: A modern loop-in rose wired up with incoming and outgoing circuit cables and a switch cable.

Right: Within BESA boxes, cable connectors link circuit cable cores to the fitting flex – on the loop-in system (bottom) or junction-box system (top).

1
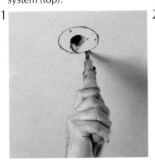

2
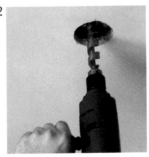

3

4
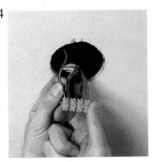

1 Mark the outline of the BESA box and use a padsaw to cut away the ceiling carefully.
2 If the box is to be screwed to the underside of a joist, use a large-diameter drill to remove just enough wood from the joist for the box to fit with its lower edge flush with the ceiling.
3 Offer up the box, checking that it fits reasonably accurately, and screw it in position.
4 Feed in the cables through the mouth of the box and use a terminal connector to link the cores as shown here and above. Note that the switch cable neutral is again flagged with red tape, and that a separate earth core links the earths in the connector with the BESA box earth terminal – necessary to provide earthing continuity with metal light fittings.

28 RUNNING CABLES

Modern electrical cable is a versatile material; you can run it almost anywhere round the house — under floors, in ceiling voids, buried in the plaster, fixed to the surface of walls and ceilings, or concealed in surface-mounted 'mini-trunking'.

Where cable is simply being fixed to wall or ceiling surfaces, it is usual to choose white-sheathed cable (and white cable clips) so that the installation looks less conspicuous against white walls or woodwork. The cable must be clipped at 250mm (10in) intervals on horizontal runs; 450mm (18in) intervals on vertical ones.

Mini-trunking makes for a neater installation when wiring is being run on the surface, and also protects the cables from accidental damage. It is often used with surface-mounted accessories, to which it is connected securely and neatly with special collars and adaptors. The trunking base is screwed, pinned or glued in position, the cable is laid in place and the cover strips are snapped on. Several sizes of trunking are available; alternatively, special skirting and architrave mouldings can be fixed in place of the existing timber ones, and used to conceal a number of cables.

Where cable is run in floor or ceiling voids, it can rest on underfloor screeds or ceiling surfaces without the need for any support. However, it should not be

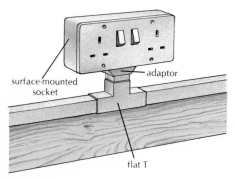
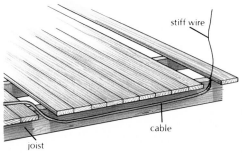

1 You can surface-mount cable on architraves and skirting boards using cable clips which have integral nails. Use white-sheathed cable to blend in with the clips and paint work.
2 Alternatively, run the cables in shallow surface-mounted PVC conduit. A snap-on cover hides the cables but allows ready access.

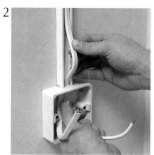

surface-mounted socket

adaptor

flat T

Mini-trunking is run along skirtings and beside architraves. Tees and adaptors link it to switches and socket outlets.

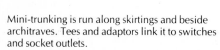

stiff wire

cable

joist

Running cables under floors is easy if you use stiff wire to draw the cable through the space underneath the floorboards.

left hanging in loops. Where the cable crosses a joist it should be threaded through a hole drilled at least 50mm (2in) below the top of the joist — *not* laid in notches cut in their tops, where floorboard nails could inadvertently damage it. In lofts, cables can cross joists, but should be routed away from loft hatches and walkways. Where polystyrene loft insulation has been laid, protect the cables with conduit or channelling; the insulation may attack and soften the cable sheathing.

Cable can be dropped down through the cavities of hollow partition walls (it's not considered good practice to run it inside exterior cavity walls, though). In stud walls (where a timber framework of vertical studs and horizontal noggins is screwed to the ceiling and floor and panelled with plasterboard), you may have to drill through the head and sole plates from the room above or below, and where there are noggins you will have to cut away a patch of plasterboard and noggin to feed the cable through. On solid walls, simply chop out a channel or chase 15 to 25mm (⅝ to 1in) deep, pin the cable in place (with conduit or channelling over it for additional protection) and plaster over it. Always run cables vertically or horizontally to and from electrical accessories, never diagonally. Then everyone knows where to expect cable if fixing things to their walls!

1 On solid walls, chop out a chase up to 25mm (1in) deep with a brick bolster and club hammer.

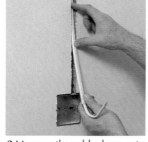

2 Measure the cable drop, cut a length of PVC conduit to match and feed the cable carefully down through it.

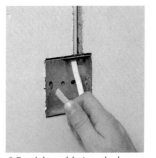

3 Feed the cable into the box via a grommet to protect it. The conduit should reach right to the edge of the box.

4 You can use small masonry pins to secure the conduit in place temporarily. Then fill the chase with plaster.

5 Where the cables have to pass across the line of floor joists, drill holes through the centres of the joists.

6 Feed the cable through the holes, leaving a little slack between joists. Never run cable in notches cut in the joists' tops.

30 EXTENDING POWER CIRCUITS

Don't be tempted to extend *any* of your power circuits if you have old rubber-sheathed cable, or you could dangerously overload an already suspect system. However, it's a comparatively easy matter to add a new socket outlet to a system with modern ring circuits. In principle, you simply add a spur (see pages 12 and 13) to one of the existing ring circuits, and you can install one spur for every socket on the original ring circuit.

You can do this in one of two ways. You can connect a branch cable into the back of an existing socket on the ring, or else cut the ring cable, insert a 30A three-terminal junction box and connect the branch cable to that. You can then run the branch cable to wherever you need the extra outlet,

and fit a single socket, a double socket or a fused connection unit. Under new wiring regulations you may have only one outlet on each spur (formerly two singles or one double).

The problem with all this is determining which cables form the ring circuit, since you aren't allowed to extend from a socket that is already on a spur, or fit a junction box on a spur cable. So far as sockets are concerned, you can eliminate two of three possibilities by unscrewing the faceplate of the socket where you want to add your spur. If you find only one cable, the socket is at the end of a spur, while if you find three cables, the socket is a ring socket with a spur already attached to it. You cannot connect a spur to either of these two sockets. If the socket

1 At a ring circuit socket (see text), connect in the new branch cable after sheathing the earth.

2 At the new socket position, fix a mounting box and draw in the new cable; note the grommet.

3 Prepare the cores and link them to the appropriate terminals. Then screw the socket faceplate into position.

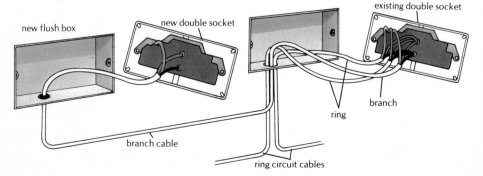

new flush box
new double socket
existing double socket
branch
ring
branch cable
ring circuit cables

has two cables it could be a ring socket, the intermediate socket on an old two-outlet spur, or an intermediate socket on an old radial circuit (there used to be restrictions on how many sockets a radial circuit could supply, but these have now been superseded — see pages 14 and 15.

Unless you are prepared to trace cables all over the house, the only way you can check whether the socket is on a ring or not is with the simple continuity tester shown below. Turn off the current at the mains. At the socket in question, disconnect the two live (red) cores from their terminals. Touch one tester lead to one core, the second to the other; if the socket is on a ring, the bulb will light, but it will stay out if the socket is on a spur or a radial circuit.

You can use the same test to find out if a cable you want to cut into is a ring or spur cable, by turning off at the mains, cutting the cable and linking the red cores as before.

If your socket is a ring socket, you can add a spur to it. That leaves the spur/radial dilemma. If you can trace the cables leaving the socket in question *back* directly to a ring socket and *on* to a socket with only one cable feeding it, then you have a spur and you must not extend from the intermediate socket. If the cables can be traced back to the consumer unit and loop on to more than one socket it's a radial circuit and you *can* extend it, provided the radial circuit serves an area of less than 20sq m (215sq ft) and is wired in 2.5mm^2 cable.

1 Cut the circuit cable and screw a three-terminal junction box to a batten between the joists.

2 Link the cable cores to the box terminals as shown. Note the earth sleeving.

3 Run in the branch cable and link its cores to the appropriate terminals as shown below.

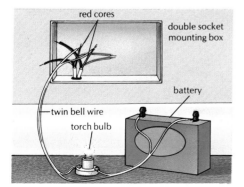

red cores
double socket mounting box
battery
twin bell wire
torch bulb

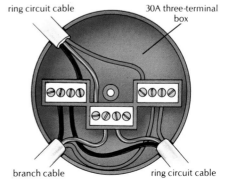

ring circuit cable
30A three-terminal box
branch cable
ring circuit cable

32 EXTENDING LIGHTING CIRCUITS

The concept of extending a lighting circuit is similar in principle to that of extending a power circuit. However, a similar caution must be given. Firstly, don't extend a circuit wired in rubber-sheathed cable. Secondly, don't overload existing circuits; you can have up to twelve 100W lighting points on each circuit, but in practice eight is a safe maximum, to allow for the use of more powerful lamps.

Start by removing the fuse (or switching off the MCB) on the circuit you want to extend and count up how many lights don't work (and how many

watts they consume). If it's less than eight lights or the total wattage is less than, say, 1,000W, then you can safely extend the circuit.

Assuming you've given the system the all-clear, you can extend it in one of two ways. Whatever system your house is wired with, you can cut the existing cable and install a four-terminal junction box, running new cables from it to the new light and switch. If you have loop-in roses with three terminal banks, you can add a spur from an existing rose without the need to use a junction box. The spur cable

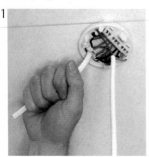
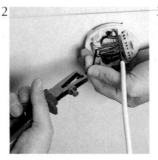
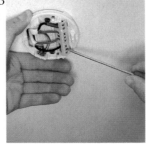

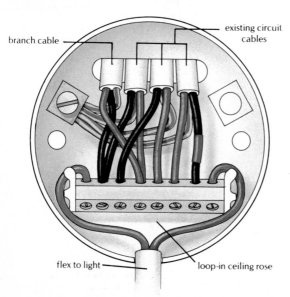

branch cable

existing circuit cables

flex to light

loop-in ceiling rose

1 To add a branch cable to a loop-in rose, unscrew the cover and draw in the new cable. A maximum of four cables can be physically connected to most loop-in roses.
2 Carefully strip the insulation from the new cable cores, and then connect them to the live, neutral and earth terminals of the existing rose.
3 At the new rose, strip the cores, sleeve the earth core and connect the cores to the appropriate terminals.

Left: With a modern loop-in rose you have room for up to four cables. Here, a branch cable to a new rose has been added alongside existing circuit and switch cables. Note that one live and one neutral terminal have to accept two cable cores each.

EXTENDING LIGHTING CIRCUITS 33

then runs on to the new rose and the new switch cable is connected to the new rose to control the new light.

Two points are worth bearing in mind when you are extending existing circuits. You may find that your circuits were wired in two-core PVC-sheathed cable . . . *without* an earth. Many types of light fitting — fluorescents and decorative pendants for instance — have metal parts which should be earthed for safety, and if they are to be installed you should run a single sheathed earth core back from the new light to the main earthing terminal at the consumer unit. It can be taken via any other ceiling roses where earthing is needed but is absent.

Secondly, you can provide power for wall lights in the same way (see pages 38 and 40). However, it may be more convenient to take a spur for your wall lights from a nearby ring main instead. This is permissible *so long as* you run the spur to a switched connection unit containing a 5A fuse (so that the lighting sub-circuit is protected by the same fusing level as an ordinary lighting circuit) before taking the cable on to the wall lights themselves.

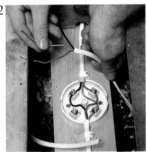
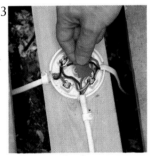

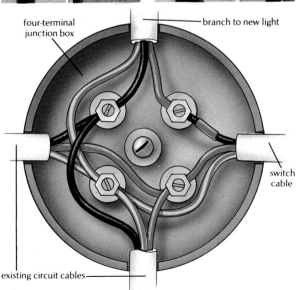

four-terminal junction box

branch to new light

switch cable

existing circuit cables

1 On a junction-box system, or if there is no conveniently-positioned loop-in rose, you can extend the lighting circuit with a four-terminal junction box. Lay the cables to the new rose and switch, and fix a batten between the joists where you want the box to be.
2 Screw the junction box to the batten. Then cut the circuit cable, prepare the cores and connect them up as shown. Prepare the ends of the new light and switch cables in the same way, sleeving the cable earth cores.
3 Connect the new cables as shown, and flag the switch cable neutral with red tape to show that it is, in fact, live.

Left: Connections for a new light and switch using a junction box.

34 **FITTING A SWITCH OR DIMMER**

Most lights are controlled by wall-mounted switches, usually sited by the entrance door and placed about 1.4m (4ft 6in) above floor level. They may be flush- or surface-mounted, and require a box 16mm (⅝in) or 25mm (1in) deep according to type. The commonest type is the plateswitch, which has one or more rocker switches in the centre of a large flat faceplate — about 85mm (3⅜in) square for one-, two- and three-gang units, and 146 × 85mm (5⅞ × 3⅜in) for four- and six-gang units. A one-way switch has two terminals marked L1 and L2 on the back, and provides on-off control from this switch position of all lights connected to it. Two-way switches have three terminals marked L1, L2 and C on the back, and are used in pairs to provide on-off control of one or more lights from two positions. Intermediate switches, with two pairs of terminals each marked L1 and L2, are used between pairs of two-way switches where control of a light or lights is needed from three (or more) positions.

Where there is no room to mount a plateswitch, you can use what is called an architrave switch instead. This is only about 32mm (1¼in) wide, and comes in one and two-gang versions (in the latter, the two rocker switches are placed one above the other).

Lights can also be controlled by ceiling switches with pull cords (and usually must be so controlled in bathrooms). These switches come in one-way and two-way versions, and can be fixed to a surface mounting box or fitted flush to a BESA box recessed into the ceiling.

Dimmer switches are light switches containing a variable control for raising

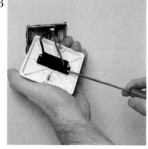

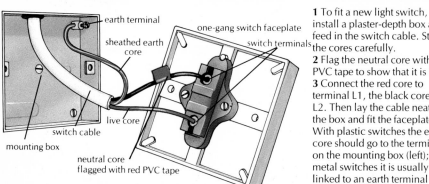

earth terminal

sheathed earth core

live core

switch cable

mounting box

neutral core flagged with red PVC tape

one-gang switch faceplate

switch terminals

1 To fit a new light switch, install a plaster-depth box and feed in the switch cable. Strip the cores carefully.
2 Flag the neutral core with red PVC tape to show that it is live.
3 Connect the red core to terminal L1, the black core to L2. Then lay the cable neatly in the box and fit the faceplate. With plastic switches the earth core should go to the terminal on the mounting box (left); with metal switches it is usually linked to an earth terminal on the faceplate itself.

FITTING A SWITCH OR DIMMER 35

or lowering the light level as well as an on-off switch. The control may be a knob, a roller or a touch-sensitive plate. Most are one-gang models, though up to four gangs are available in some ranges. When fitting one in place of an existing plateswitch, you may have to fit a deeper box (or even a two-gang box in the case of multi-gang dimmers) but the connections are usually the same as for ordinary switches.

A point to remember when fitting new plateswitches in place of existing ones for purely cosmetic reasons is to save the old fixing screws. The screws supplied with the new switch may not thread into the lugs on the old box if this pre-dates the introduction of metrication.

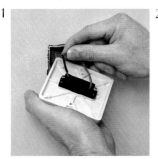
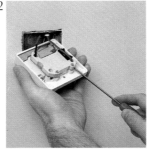

1 To fit a one-gang dimmer (left) in place of an existing one-way switch, first remove the faceplate.
2 Connect the cores to the dimmer terminals, following the maker's instructions carefully. Below: with two-gang dimmers, two switch cables will be present.
3 Fix the dimmer to the box. With this model a push-on knob covers the fixing screws.

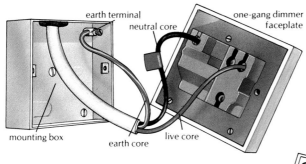

earth terminal
neutral core
one-gang dimmer faceplate
mounting box
earth core
live core

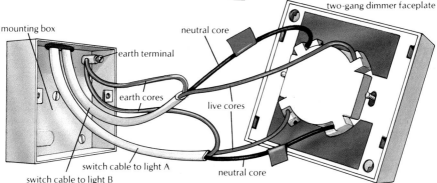

two-gang dimmer faceplate
mounting box
neutral core
earth terminal
earth cores
live cores
switch cable to light A
switch cable to light B
neutral core

36 TWO-WAY SWITCHING

It's often convenient to be able to switch a particular light on or off from more than one place in the house — for example, to control the hall light from hall or landing, or a bedroom light from near the door or bedside. So instead of having a single one-way switch breaking the live core of the mains supply, you use two two-way switches, one at each switch position. Each switch has three terminals, labelled L1, L2 and C; the live supply goes to the C terminal of one switch while the C terminal of the other switch is linked to the light being controlled. The pairs of L1 and L2 terminals are linked by strapping wires — L1 to L1 and L2 to L2 — so that, whatever the position of one switch,

the light can still be controlled from the other. That's the theory.

In practice, things are a little more complicated, since special 1.0mm² three-core and earth cable is used. This has cores colour-coded in red, blue and yellow (plus the regular earth core); red should always be used for the C to C link, yellow for the L1 to L1 link and blue for the L2 to L2 link between the two switches. The switch 'drop' cable from the junction box or loop-in rose on the lighting circuit is run to the nearest two-way switch, its red core being connected to L1 and its black core to L2. The earth core of the three-core and earth cable links the earth terminals of the two switches, so

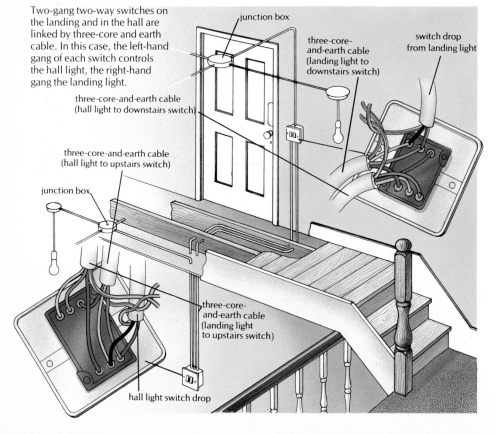

Two-gang two-way switches on the landing and in the hall are linked by three-core and earth cable. In this case, the left-hand gang of each switch controls the hall light, the right-hand gang the landing light.

junction box

three-core-and-earth cable (landing light to downstairs switch)

switch drop from landing light

three-core-and-earth cable (hall light to downstairs switch)

three-core-and-earth cable (hall light to upstairs switch)

junction box

three-core-and-earth cable (landing light to upstairs switch)

hall light switch drop

TWO-WAY SWITCHING 37

providing earth continuity throughout the switching circuit. The earth core must, of course, be sleeved with green and yellow PVC when exposed.

So to convert control of a light to two-way switching you simply replace the existing one-way switch with a two-way unit, and run three-core and earth cable on to the new switch position. You can then provide further intermediate switches between the two-way switches at any point.

If you have partial two-way switching — say, you can switch the landing light on from hall or landing, but the hall light from the hall only — you follow a similar sequence. Replace the one-gang one-way switch on the land-

ing with a two-gang two-way switch and run a second three-core and earth cable alongside the one already linking the two switches. One gang of each switch is then linked by three-core cable, so enabling you to switch each light from upstairs or downstairs.

Lastly, you may want two-way switching of more than two lights — in a bedroom, say, with two independently-switched bedside lights. This could obviously involve a great deal of cable, but you can avoid a lot of the extra work by fitting a multi-terminal junction box in the ceiling void and making your cable connections through this. One typical wiring arrangement using this system is shown below.

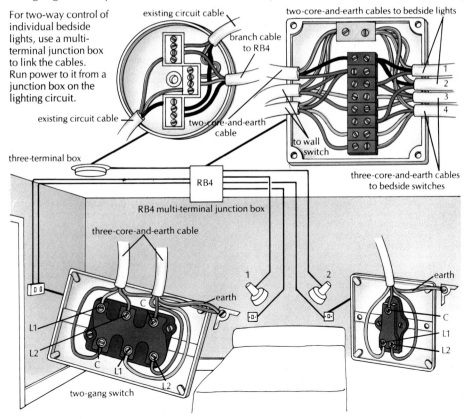

For two-way control of individual bedside lights, use a multi-terminal junction box to link the cables. Run power to it from a junction box on the lighting circuit.

existing circuit cable

branch cable to RB4

two-core-and-earth cables to bedside lights

existing circuit cable

two-core-and-earth cable

to wall switch

1
2
3
4

three-terminal box

RB4

three-core-and-earth cables to bedside switches

RB4 multi-terminal junction box

three-core-and-earth cable

earth

1

2

earth

C

L1

L2

C

L1

L2

C

L1

L2

two-gang switch

38 MOVING OR ADDING A PENDANT LIGHT

It's a relatively simple job to move an existing pendant light to a new position, especially if the switch is to be left where it was, or to add a new light to an existing circuit. In the first case, what you do depends on what you find under the cover of the old rose.

If you find only one cable present, it's wired on the junction box system and all you have to do is mount a suitable junction box to a joist above the old light position, connect the existing cable to its terminals and then run a new cable from the box to the new light position (shown below and top right).

If you find two (or more) cables at the old rose, it's wired on the loop-in system, and in this case you need a four-terminal junction box to which the existing cables, plus the spur to the new light, are connected (right). In either case, connect the new spur to the new rose and make good the ceiling where the old rose was removed.

In the case of loop-in wiring, you could choose instead to leave the existing rose where it is (but disconnect the pendant flex, leaving the option open of replacing it in the future should the lighting point be needed again) and to loop-in a spur cable from this rose on to the new rose position.

To install a completely new light, a little more work is involved. First of all, you have to check out how many lights are already on the circuit. You then have to lift floorboards or go into the

1 Disconnect the flex and cable from the old rose, and draw the cable into the ceiling void.

2 Connect the existing and new cables to a junction box mounted on a batten.

3 Fit the cover on the box. Then remove the old rose from the ceiling below and make good.

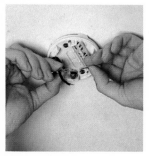

4 Mount the old rose in the new position, feed in the new cable and sleeve the earth core.

5 Connect the cores of the light flex to the rose terminals; hook them over their hangers.

6 Connect the other end of the flex to the lampholder after threading it through its cover.

MOVING OR ADDING A PENDANT LIGHT 39

attic to locate a circuit cable or loop-in rose near to your proposed light position. Make sure any cable *is* a circuit cable, *not* a switch cable: it must link two junction boxes or roses. If a circuit cable is nearby, isolate it at the consumer unit, cut it and insert a four-terminal junction box. Then run new cables from the box to the new rose and the new switch (remember that if you want the new switch at an existing switch position, you can run the new switch cable to that position and simply fit a new plateswitch there with one extra gang). If a loop-in rose is nearest, add a spur to the rose terminals, run it to a new rose at the new light position and then run a new switch cable from the new rose to the new switch.

Below: An alternative way of adding an extra light is to connect it to an existing light and switch which then controls both lights.

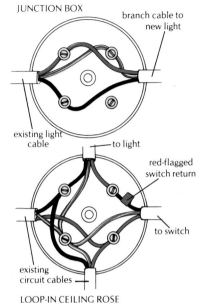

JUNCTION BOX

branch cable to new light

existing light cable

to light

red-flagged switch return

to switch

existing circuit cables

LOOP-IN CEILING ROSE

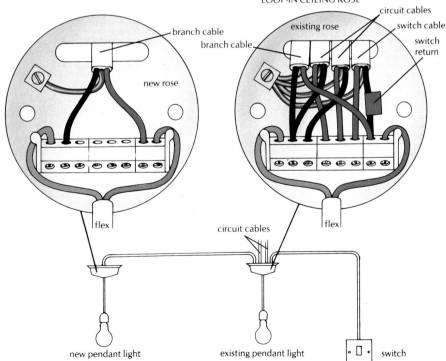

branch cable

branch cable

new rose

circuit cables

existing rose

switch cable

switch return

flex

flex

circuit cables

new pendant light

existing pendant light

switch

40 FITTING WALL LIGHTS

Wall lights help to add variety to a lighting scheme, providing localised illumination that can be decorative or functional, depending on what type of fitting you choose (and there is an enormous range available). Installing whatever light you finally decide on is a comparatively simple matter, but the job will make a mess of your decorations unless you are prepared to put up with surface-mounted cable, since cables have to be chased down walls to the light and any new switches required.

Mounting the light on the wall is the easy part. The fitting must be screwed over a flush-mounted enclosure that houses the connector block linking the fitting flex to the circuit cable; this can be a circular BESA box — ideal for fittings with circular base-plates, since the fitting can often be screwed direct to the box's fixing lugs — or an architrave box if the fitting has a very narrow baseplate (in this case, the fitting will probably be mounted via screws and wallplugs driven into the wall).

Next, you have to decide how to wire up the lights. You can either feed them from a convenient lighting circuit, as described on pages 38 and 39

1 To fit a wall light with its own separate switch, channel out the wall and fit a BESA box. Continue the chase on down to the switch position.

2 Feed the switch cable down to the BESA box, and then on down to the position of the switch. Pull it through so it lies neatly in the base of the box.

3 Next, feed in a second cable to supply the light fitting itself. The two are linked at a four-terminal junction box in the ceiling void above.

4 Strip the cable cores and then connect them to a three-terminal insulated connector block as shown.

5 Strip the cores of flex attached to the light fitting (three-core flex for a metal fitting) and connect them to the terminals.

6 Screw the fitting to the box with machine screws. If the fitting has a large baseplate, screw it to the wall instead.

FITTING WALL LIGHTS 41

(using three-terminal junction boxes to split the feed cables to two lights), or from a ring main. The latter method is particularly convenient in ground floor rooms, since it avoids having to lift first floor floorboards to make the circuit connections.

You should be able to pick up power from a ring main in two ways; the first involves cutting a nearby ring circuit cable, inserting a 30A three-terminal junction box and running a spur from there; in the second case, you simply run the spur cable from a conveniently-sited ring main socket (*not* a spur socket — see pages 30 and 31). In either case, the spur must be in 2.5mm^2 cable, and must run directly to a switched fused connection unit containing a 5A fuse, so that the new lighting circuit is protected by a smaller fuse than the 30A ring circuit fuse, and can be isolated if necessary for maintenance or alterations.

You can then run a length of 1mm^2 cable on from the fused connection unit, either direct to the light (in which case, the fused connection unit acts as the light switch, too), or to a four-terminal junction box from where cables run on to the light and to a separate switch position.

1   2

1 If the light fitting has an integral switch you can mount it over a BESA box or . . .
2 . . . an architrave box. Note that in each case the box is connected to the cable earth.
Below: You can supply wall lights from an existing loop-in rose (A) or from a three-terminal junction box (B). A new four-terminal junction box provides switch connections.

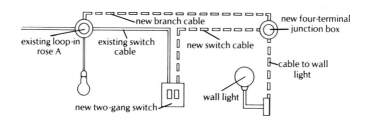

existing loop-in rose A — new branch cable — new four-terminal junction box
existing switch cable — new switch cable — cable to wall light
new two-gang switch — wall light

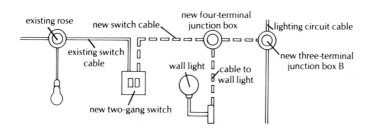

existing rose — new switch cable — new four-terminal junction box — lighting circuit cable
existing switch cable — wall light — cable to wall light — new three-terminal junction box B
new two-gang switch

42 FITTING DECORATIVE LIGHTS

You don't have to restrict yourself to having pendant lights at all your ceiling fittings. The central light in many rooms is now usually supplemented by wall lights, table and standard lamps, and so on. However, it still has an important role to play, even if the tendency is towards fittings that are decorative rather than purely functional. You can, of course, simply change lampshades to transform an ordinary pendant fitting, but much more attractive effects can be obtained with other types of decorative fitting.

Such fittings usually either hang down from the ceiling on a rod, chains or cords, or else are close-mounted on the ceiling surface. What's involved in installing them varies from fitting to fitting, but one or two general guidelines can be given. With pendant-type fittings you will usually have to install a flush-mounted BESA box, screwed securely to a batten so it can carry the weight of the fitting screwed to it. The box then houses the connector blocks that link the fitting flex and the mains cable — you'll need four connectors instead of three if you're replacing an existing loop-in rose.

With surface-mounted fittings, you often have to mount a fixing bar on the

1 Most decorative light fittings are mounted over a BESA box. Connect the cable cores to a four-terminal connector block on a loop-in wiring system.

2 Follow the fitting manufacturer's wiring instructions. Then fit the block neatly into the box and fix the fitting baseplate to it.

3 With this downlighter, all that remains is to offer up the cowl, locking it on to the lugs on the baseplate by twisting it slightly, and fit the lamp.

1 If the fitting does not need to be earthed, link the block to the box with a short length of sleeved earth wire.

2 As before, offer up the new fitting and connect the flex cores to the terminals of the connector block.

3 Finally, position the fitting baseplate accurately over the BESA box, and tighten up the fixing screws fully.

FITTING DECORATIVE LIGHTS 43

ceiling, screw to it a base plate carrying the lampholder and terminal block, make the connections (if there are only three terminals, you can't handle loop-in connections) and fit the lamp cover. With rise-and-fall fittings, a base plate and hook (for the rise-and-fall unit) are usually screwed to a BESA box and concealed by a plastic cover.

Another versatile form of decorative lighting is the track fitting, which is surface-mounted on the ceiling and carries a number of sliding spotlights. Electrical conductors along the length of the track allow the spotlight position to be adjusted at will. There are no facilities for loop-in wiring with track lighting, so if you have such a circuit you will have to break into it with a four-terminal junction box and run cables to the track and its switch.

Take care when installing decorative fittings to follow the manufacturer's instructions about the maximum bulb wattage to be used. In the case of enclosed fittings, it's also a good idea to slip lengths of heat-resistant sleeving over the flex and cable cores before connecting them to the terminal block; otherwise, the build-up of heat inside the enclosed fitting could make the insulation brittle and eventually fail.

1 With rise-and-fall fittings, link the cable cores to a connector block. Then attach the base plate and hook to the BESA box with machine screws.

2 Next, hook the rise-and-fall unit over the hook. You can adjust the up-and-down movement simply by tightening or loosening the control screw.

3 Connect the flex cores from the rise-and-fall unit to the correct terminals of the connector block.

4 Push the plastic cover up to the ceiling to conceal the rise-and-fall unit and secure it with a grub screw.

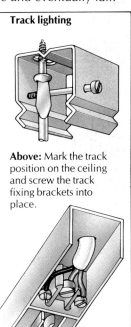

Track lighting

Above: Mark the track position on the ceiling and screw the track fixing brackets into place.

Above: Thread in the feed cable, attach the track to the brackets and connect the cores to the terminals.

44 INSTALLING A FLUORESCENT FITTING

Many people have mixed feelings about fluorescent lights because of the quality of the light they emit — harsh and bright. However, they need not look clinical; they are an ideal way of providing good all-round lighting in rooms such as kitchens, bathrooms and workshops, and can be used to provide all sorts of concealed lighting effects when fixed behind baffles and cornices, or above worktops.

Fluorescent lights cost more to install than ordinary pendant lights, but are more economical to run because they are more efficient. A 1500mm (5ft) tube rated at 65W gives out four times as much light as a 100W tungsten filament lamp.

A fluorescent fitting is installed in much the same way as any close-mounted ceiling fitting. It usually comes as an integral fitting in two parts; the base plate carries the two lampholders (one at each end of the tube), the control gear (to start the discharge through the tube and keep it running when the light is on) and the terminal block that links the circuit cable to the fitting. Over this is fitted a cover or 'diffuser'. You can also buy the lampholder and control gear separately — useful when you want to mount several

1 Remove the diffuser and tube from the fitting and take off the backplate cover by undoing the retaining screws and set it to one side.

2 Offer up the baseplate to the ceiling and mark the positions of the fixing screws through the holes in the backplate. Screw into a joist or a batten.

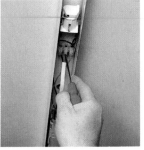

3 Feed in the supply cable, strip the cores and connect them to the appropriate terminal of the connector block on the baseplate of the fitting.

4 If using existing cables that aren't earthed, add an earth core between the fitting and the consumer unit (see page 33).

5 Offer up the baseplate cover, fitting it carefully between the ends of the unit, and do up the retaining screws.

6 Fit the tube and diffuser. In this case, the diffuser is held on to the tube by spring clips, so is actually fitted first.

INSTALLING A FLUORESCENT FITTING 45

tubes behind a pelmet to provide perimeter lighting, or below kitchen wall units to light a length of worktop, since one set of control gear can be linked to several tubes.

Incidentally, it is possible to dim fluorescent lighting; but you need to use a special model of dimmer switch made for the purpose.

When connecting up a fluorescent fitting in place of an existing rose, you should follow the same sequence of operations for wiring it as that given for pendant lights on pages 38 and 39 — what you do depends on the wiring at the existing rose.

Below: Connect the control units to a three-terminal junction box (A). With junction-box wiring, run new cable from the existing four-terminal box (B). With loop-in wiring, replace the flex to the existing fitting with new cable (C).

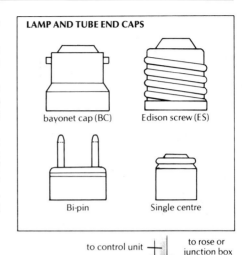

LAMP AND TUBE END CAPS

bayonet cap (BC)   Edison screw (ES)

Bi-pin   Single centre

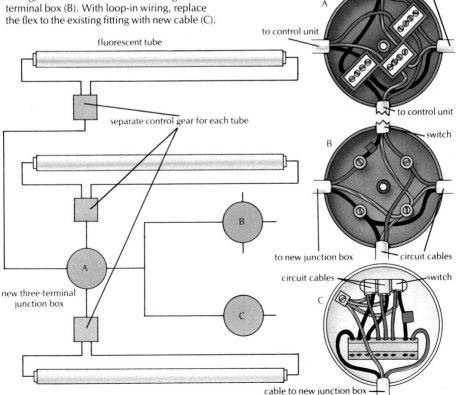

fluorescent tube

separate control gear for each tube

new three-terminal junction box

to control unit   to rose or junction box

A

to control unit

to control unit

switch

B

to new junction box   circuit cables

circuit cables   switch

C

cable to new junction box

46 CHANGING SINGLE SOCKETS INTO DOUBLE

One of the simplest ways of getting extra power points around the house is to convert existing single sockets into double ones. Only do this if your house is wired in PVC-sheathed cable; you should not risk overloading any circuits wired in rubber-sheathed cable.

You can carry out the job in several ways, and which you choose depends on whether you have surface- or flush-mounted sockets already, and whether you want your new sockets to be flush- or surface-mounted.

The simplest option involves fitting surface-mounted sockets. First isolate the circuit, unscrew the existing socket faceplate and disconnect the cable from the terminals. If the existing socket is surface-mounted, unscrew the one-gang mounting box, fit a two-gang surface-mounted box in its place, connect the cable to the new double faceplate and screw it to the mounting box. If the existing socket is flush-mounted, leave the box where it is and screw the new two-gang surface-mounted box over it (holes in its rear face will line up with the lugs on the existing box) before adding the faceplate as before.

You may want your new double socket to be flush-mounted. In this case, disconnect the existing faceplate and remove the existing mounting box. If this was surface-mounted, cut a recess for a two-gang metal mounting box; if it was flush-mounted, simply enlarge the existing one-gang hole to take a two-gang box. Then secure the new box (having threaded the cable into it through a knockout fitted with a rubber grommet), connect the cable to the terminals and fit the new double faceplate to the box. Make good any gaps round the perimeter of the new box with plaster or filler.

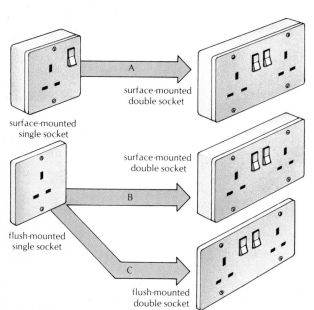

surface-mounted
single socket

surface-mounted
double socket

A

flush-mounted
single socket

surface-mounted
double socket

B

flush-mounted
double socket

C

Left: The simplest option is to change a surface-mounted single socket into a surface-mounted double one. Flush single sockets can be converted into surface-mounted or flush doubles quite easily, but in the latter case the wall recess has to be enlarged to accept the new, larger socket.

Below: The three options, shown in cross-section.

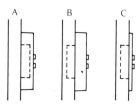

A   B   C

CHANGING SINGLE SOCKETS INTO DOUBLE 47

1 To change a surface-mounted single socket into a double, unscrew and disconnect the faceplate, then unscrew the mounting box from the wall. Lift it away carefully.

2 Remove a knockout from the base of the new double mounting box, thread in the cables and mark the fixing screw positions. After drilling and plugging, fit the box.

3 Finish off by connecting the cable cores to the appropriate terminals of the new double socket faceplate. Press the cables back neatly and fix the faceplate in position.

1 To change a flush single socket into a surface-mounted double, unscrew the faceplate of the single socket and disconnect the cable cores. Set the faceplate aside.

2 Remove a knockout from the base of the new double mounting box, and thread in the cables. Check which of the fixing screw holes coincide with the old box lugs.

3 Use the screws that secured the old faceplate to fix the new box in position over the existing one. This will ensure the threads match. Check the new box is level as you tighten the screws.

1 To convert a flush single socket to a flush double, remove the old faceplate and box. Chop out a larger recess.

2 Offer up the new box, thread in the cables and mark the positions of the fixing screws. Drill and plug the holes.

3 Fix the mounting box in place. Then connect the cable to the faceplate terminals and attach it to the box.

48 FITTING A NEW SOCKET

When you decide to fit a new socket outlet where none exists, you have two separate jobs to tackle. The first is to run a power supply to the new socket position (see pages 30 and 31), and the second is to fix a mounting box of whatever type you prefer to the wall, ready to receive the new cable and faceplate (see pages 24 and 25).

Most socket outlets need a mounting box 35mm (1⅜in) deep, but where you are fitting flush boxes in cavity and single brick internal walls, chopping such a deep hole can be difficult without cutting right through the wall, so shallow 25mm (1in) boxes are used instead, coupled with accessories having thicker-than-usual faceplates.

Remember that you can also fit a special dual box instead of a two-gang box, and use two different one-gang faceplates — say, one single socket and one fused connection unit taking power to a freezer. You can combine any two accessories in this way, with one exception: you are *not* allowed to house a TV coaxial socket outlet in the same box as mains exceeding 50 volts; so you can't put one alongside the socket outlet powering the TV set; the Wiring Regulations forbid it.

With solid walls, making fixings is

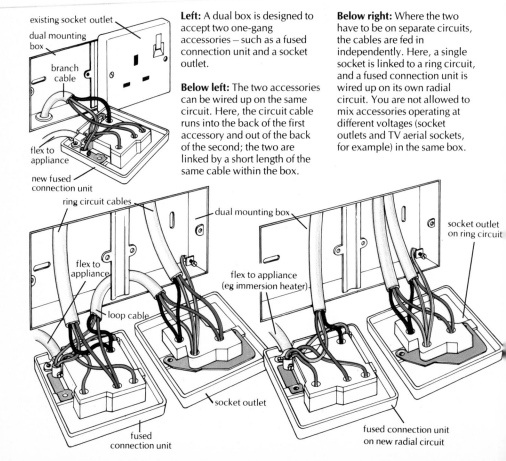

existing socket outlet
dual mounting box
branch cable
flex to appliance
new fused connection unit

Left: A dual box is designed to accept two one-gang accessories – such as a fused connection unit and a socket outlet.

Below left: The two accessories can be wired up on the same circuit. Here, the circuit cable runs into the back of the first accessory and out of the back of the second; the two are linked by a short length of the same cable within the box.

Below right: Where the two have to be on separate circuits, the cables are fed in independently. Here, a single socket is linked to a ring circuit, and a fused connection unit is wired up on its own radial circuit. You are not allowed to mix accessories operating at different voltages (socket outlets and TV aerial sockets, for example) in the same box.

ring circuit cables
flex to appliance
loop cable
dual mounting box
flex to appliance (eg immersion heater)
socket outlet on ring circuit
socket outlet
fused connection unit
fused connection unit on new radial circuit

FITTING A NEW SOCKET 49

obviously no problem: you simply drill holes where you need them, insert wallplugs and drive in the fixing screws, but with hollow stud partition walls life is not so simple. Obviously, if you are building the wall from scratch, you can fix the mounting boxes to noggins between the studs and cut holes in the plasterboard cladding ready for flush accessories to be fitted later. With existing walls, however, you have two choices. You can either use surface-mounted fittings (screwed to studs if they're conveniently placed, or secured with appropriate cavity fixings) or cut holes in the plasterboard

and install a flush box. This can be done in two ways.

One rather fiddly method involves clipping metal lugs to each side of the metal box and offering it up to the inner face of the plasterboard; you need string to pull it tightly into place while you attach the socket faceplate, but once the fixing screws are tightened the whole assembly is quite secure. The other method involves the use of 'flange boxes', which are set into the hole and secured at each corner with screws and cavity fixings. The socket faceplate can then be mounted on the flange box.

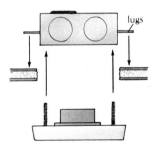

1 To mount accessories in hollow partition walls, clip small lugs to the mounting box. These are drawn against the inside of the partition as the faceplate is screwed on.

2 Cut away the plasterboard to match the size of the mounting box, drilling a hole in each corner and then cutting between the holes with a padsaw. Then fit the lugs to the box.

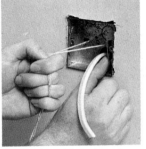

3 Draw the cable in through a knockout and feed a loop of string through the box base. Offer it up inside the recess and pull the string to hold it in place.

4 Holding the string taut, offer up the accessory faceplate and start to turn the fixing screws into the box lugs.

5 When you have almost fully tightened the screws, release the string loop and draw it out. Finish tightening the screws.

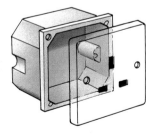

6 To mount accessories in holes cut in thin sheet materials, use a special flanged mounting box of the appropriate size.

50 FUSED CONNECTION UNITS

Fused connection units are mainly used instead of plugs and sockets to link certain electrical appliances — fridges and freezers, for instance — to the mains when you want to avoid the risk of the appliance being unplugged accidentally. They are more convenient, too, for items that are often in use — such as washing machines, extractor fans and waste disposal units — because they save you the bother of constantly plugging in and unplugging the flex.

A connection unit has a double-pole switch (one that breaks both the live and neutral cores of the circuit to the appliance) and a cartridge fuse that acts in the same way as a plug fuse to protect the flex or cable running on from the unit. Different types of unit are

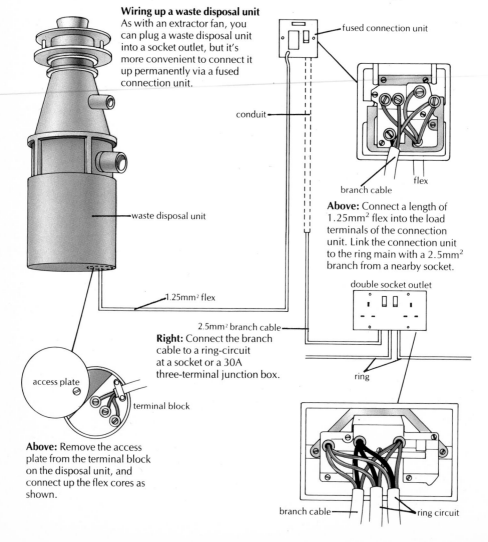

Wiring up a waste disposal unit
As with an extractor fan, you can plug a waste disposal unit into a socket outlet, but it's more convenient to connect it up permanently via a fused connection unit.

fused connection unit

conduit

waste disposal unit

branch cable

flex

Above: Connect a length of 1.25mm² flex into the load terminals of the connection unit. Link the connection unit to the ring main with a 2.5mm² branch from a nearby socket.

double socket outlet

1.25mm² flex

2.5mm² branch cable

Right: Connect the branch cable to a ring-circuit at a socket or a 30A three-terminal junction box.

ring

access plate

terminal block

Above: Remove the access plate from the terminal block on the disposal unit, and connect up the flex cores as shown.

branch cable

ring circuit

FUSED CONNECTION UNITS 51

available, with or without switches or neon indicators; some are designed for use in spur circuits, with cable entering and leaving the unit (as in the wall light installation described on pages 40 and 41), while others are designed to allow flex to run through an outlet in the face of the unit to the appliance.

Connection units are one-gang units and are fitted to 25mm (1in) or 35mm (1⅜in) deep mounting boxes. Slim and standard depth versions are available as for socket outlets. They can be mounted in pairs (or alongside a single socket outlet) in dual boxes.

A particular type of low-rated unit is the clock connector, containing a 2A fuse and intended to provide a permanent connection for mains clocks and other low-rated appliances.

Fitting an extractor fan

There's nothing to stop you simply plugging an extractor fan into a nearby socket outlet, but a far neater solution is to wire it in permanently.

new three-terminal junction box

power supply

cable in conduit

to clock connector

power supply

Above: Provide the power for the fan via a branch cable taken from a nearby lighting circuit. Use a junction box or connect into a loop-in ceiling rose.

Left: Connect the flex from the clock connector to the fan's terminal block as shown.

terminal block

three-core flex

back of fuse unit

Above: Connect the flex to the plug-in part of the clock connector and fit the fuse.

flex

fan

fused clock connector

Right: Run 1mm² cable from the junction box to the fixed part of the clock connector. Connect in the switch cable too, with a cable connector.

switch cable

one-gang switch

Left: At the switch, flag the black core with red PVC tape to denote that the core is, in fact, live.

52 ADDING A NEW CIRCUIT

Unless your electrical system was extremely well-planned at the very beginning, it's quite likely that sooner or later you will want to add one or more new circuits. For example, you may have built an extension that requires its own ring main, or you may want to add circuits to such major appliances as an electric cooker, an instantaneous shower or an immersion heater, all of which must be fed from an individual radial circuit.

In a modern installation, you may be lucky enough to have a consumer unit with a number of unused fuseways that could be used to supply these circuits, but in most cases you will have to install extra circuit fuse capacity to power the new circuits. This involves installing either a second consumer unit alongside the first (worthwhile if you want several extra circuits; you could even consider removing the existing unit and fitting a much larger one to handle both existing and new circuits) or a smaller unit called a

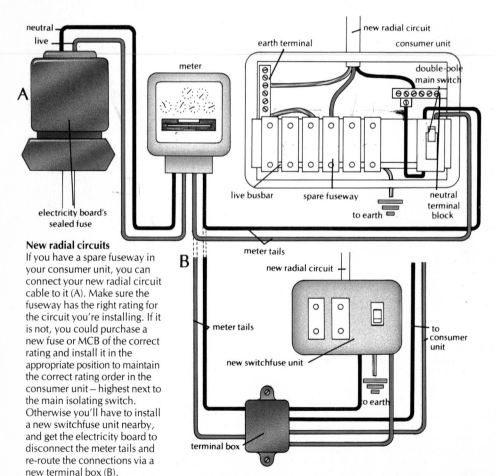

A

neutral
live

electricity board's
sealed fuse

meter

new radial circuit

consumer unit

earth terminal

double-pole
main switch

live busbar   spare fuseway

to earth

neutral
terminal
block

New radial circuits
If you have a spare fuseway in your consumer unit, you can connect your new radial circuit cable to it (A). Make sure the fuseway has the right rating for the circuit you're installing. If it is not, you could purchase a new fuse or MCB of the correct rating and install it in the appropriate position to maintain the correct rating order in the consumer unit – highest next to the main isolating switch. Otherwise you'll have to install a new switchfuse unit nearby, and get the electricity board to disconnect the meter tails and re-route the connections via a new terminal box (B).

B

meter tails

new radial circuit

meter tails

new switchfuse unit

to earth

to
consumer
unit

terminal box

ADDING A NEW CIRCUIT 53

1 When you want to add a new circuit, check at your consumer unit to see if you have any spare fuseways or room within the unit to insert another fuseway. If you have room to add one, switch off the main switch on the unit and undo the fixing screws holding the MCBs or fuseholders to the neutral busbar. This unit has room for two more circuits to be added.

2 Where the new MCB or fuseholder is fitted depends on the current rating of the circuit you want to add. In this case the new circuit is a 15A one, and so an MCB of the appropriate rating has to be inserted between the low-rated lighting circuit MCBs furthest from the main switch and the higher-rated MCBs nearer to it. This order of circuit rating must be followed.

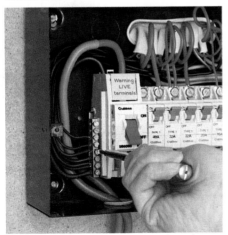

3 With the new MCB or fuseholder clipped into place, the circuit cable can be fed into the consumer unit. Remove enough of the cable's outer sheathing to allow all the cores to reach the appropriate terminals within the unit. You can always cut the cores down if they are too long. Then connect up the cores; here, the neutral is being connected to the neutral block.

4 Next, the earth core, sleeved in green/yellow PVC, is connected to the unit's earth terminal (top right). Then, the live core is linked to the top terminal of the new MCB or fuseholder to complete the connection of the new circuit cable. All that remains is to fit the consumer unit's protective cover back in place and turn on the main switch and circuit MCBs.

54 ADDING A NEW CIRCUIT

mainswitch and fuse unit, which incorporates an isolating switch and usually one or two extra fuseways. This is mounted alongside the original consumer unit or fusebox. The meter tails are then disconnected from the existing unit and taken on to a terminal box; from there, two new sets of single-core double-insulated cables run to the existing unit and the new one. You can do all the installation work on the new unit, but you must leave the disconnection

1 If your consumer unit doesn't have any spare fuseways, put in a new switchfuse unit.
Attach its baseplate to the board near the old box.

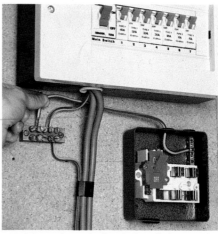

2 Link its earth terminal to the installation's main earthing point with a single-core earth lead of the appropriate size – usually 6mm².

3 Connect in live and neutral single-core tails to link the new switchfuse unit with a terminal box. This is necessary because only one pair of tails can be connected to the meter.

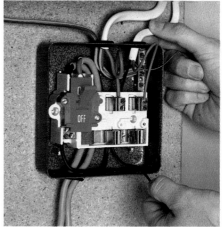

4 You can now feed in the cables that will take power to whatever new circuits you are installing. Here, the neutral cores are cut long enough to pass behind the fuseholders.

of the meter tails, and their reconnection to the new box, to the local electricity board.

A new switchfuse unit must also be properly earthed, which means that a length of single-core earth cable must be taken from the new unit's earth terminal to the installation's main earthing point. If you are not sure where this is, ask the electricity board to make the connection when they reconnect the meter tails.

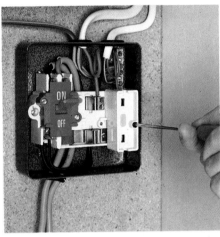

5 Now the fuseholders can be screwed into position over the terminals. Check that they sit squarely in place before tightening up the screw.

6 Fit the fuses or MCBs into the fuseholders, check that all connections are neat and secure, then replace the cover of the switchfuse unit.

7 Take the tails from the new unit to a terminal box mounted nearby. Then call in the electricity board to disconnect the consumer unit from the meter and connect up the new installation.

8 The completed installation looks like this. Note how the consumer unit and switchfuse unit are linked to the terminal box, from where one set of meter tails lead to the meter.

56 COOKER CONTROLS

Because cookers are heavy current users, they must be wired on a separate radial circuit run from the consumer unit. Its rating will depend on the wattage of the cooker — one rated at up to 11kW should be on a circuit wired in 6mm² cable and protected by a 30A fuse, while one with a higher rating should be wired with 10mm² cable on a 45A fuse. At the cooker end

of the circuit, what you do depends on the type of cooker you have.

With a freestanding type of cooker, the most common arrangement is to have a cooker control unit — an accessory combining a double-pole switch and a 13A socket outlet — or else just a double-pole switch (usually marked 'cooker'). The circuit cable runs to this switch unit (which must be

Wiring up a separate hob and oven
Right: If you want the control unit between the hob and oven, run separate 6mm² cables to each from the control unit. Each appliance must be no more than 2m (6ft 6in) from it.

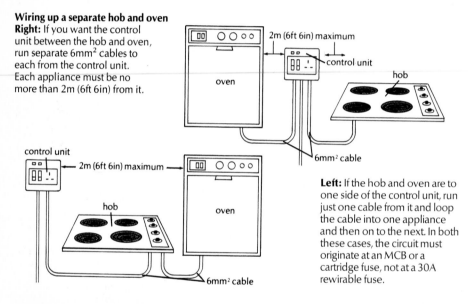

2m (6ft 6in) maximum

control unit

oven

hob

6mm² cable

control unit

2m (6ft 6in) maximum

oven

hob

6mm² cable

Left: If the hob and oven are to one side of the control unit, run just one cable from it and loop the cable into one appliance and then on to the next. In both these cases, the circuit must originate at an MCB or a cartridge fuse, not at a 30A rewirable fuse.

1 Feed the circuit cable and cooker cable into the mounting box. Secure the cooker cable in the clamp provided.

2 Prepare the cable ends, then connect the cores of each cable to the appropriate terminals on the control unit's faceplate.

3 Check the connections. Then snap on the conduit cover strips to conceal the cables, and fix the faceplate in position.

COOKER CONTROLS 57

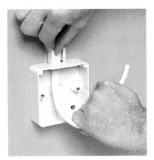

1 The cable from the control unit passes on to a cooker connection unit. Feed the cable into the box through a knockout. Cut off any excess cable with wire cutters.

2 Connect the cable that will run to the cooker into the lower set of terminals, and secure it in the cable clamp. Tighten the fixing screws fully to prevent the cable from being pulled out.

3 Now connect the cores of the supply cable from the control unit into the other set of terminals. Part the conductors if necessary to fit them under the terminal clamp.

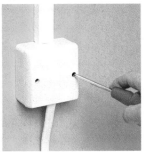

4 Now snap on the conduit cover, and attach the connection unit's cover. Note that a cooker is the only appliance linked to the mains with cable instead of flex.

5 Feed the other end of the cable from the connection unit into the rear of the cooker. Make sure you use the cable clamp; follow the maker's instructions for connections.

6 Check that all of the connections are secure and that the cable earth core has been sleeved. Then tighten up the cable clamp and the cooker is ready to be switched on.

mounted within reach of the cooker) and then a further cable is run on to a connector unit which is fixed lower down the wall.

From here, cable (not flex; the Wiring Regulations allow this unusual exception) runs to the cooker's terminal block; this means that when repairs are necessary, the cooker circuit can be turned off at the double-pole switch and the cooker can be disconnected at the connector unit without having to disturb the fixed wiring.

With split-level oven and hob units, both can be linked directly to one cooker control unit or switch unit if they are both within 2m (6ft 6in) of it. Where the switch is between the oven and hob, two cables are run from the switch to the two units; where it is to one side, one cable is looped into the nearer appliance, and then runs on to the further one.

Cooker control units and double-pole switches can be flush- or surface-mounted; the former need a 55mm (2¼in) deep box, while the latter come complete with a special metal box. Cooker connector units are normally flush-mounted.

58 INSTANTANEOUS SHOWERS

Instantaneous electric showers take water direct from the mains supply, pass it over powerful electrical heating elements and supply it to a shower rose, which may be installed over a bath or shower tray.

Early models provided temperature control by varying the flow of water through the heater, but they were something of a disaster because their throughput of water was very low and the temperature was likely to be affected by any sudden draw-off from taps elsewhere in the system.

Recent models are more powerful (rating up to 8kW) and incorporate temperature stabilisers which iron out the problem of pressure fluctuations.

The power supply for these showers must be supplied from a separate radial circuit, wired in 6mm² cable, and the unit itself must be controlled by a 30A or 45A double-pole cord-operated switch. This can be surface-mounted on a plastic box fixed to the ceiling, or flush-mounted on a BESA box recessed into it and screwed to a batten fixed between the joists. This type of switch also incorporates a neon light to indicate when the unit is switched on. A wall-mounted double-pole switch can be used but must be outside the room.

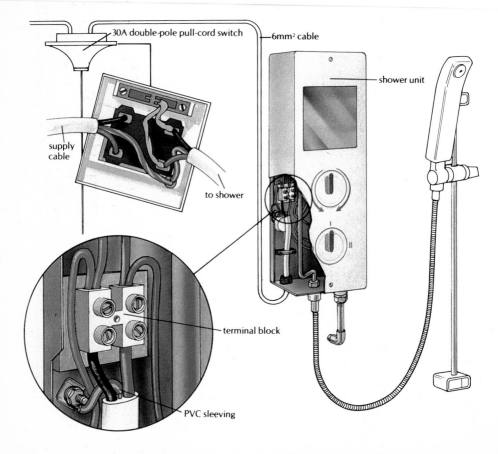

30A double-pole pull-cord switch

6mm² cable

shower unit

supply cable

to shower

terminal block

PVC sleeving

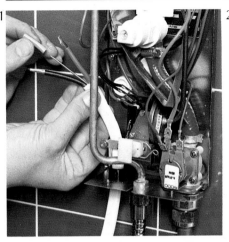

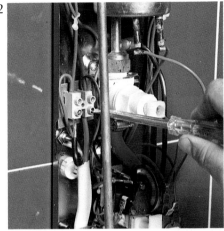

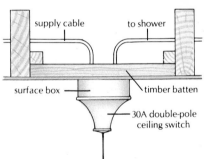

supply cable — to shower

surface box — timber batten

30A double-pole
ceiling switch

Above: Because of the pull exerted on a heavy-duty ceiling switch like this, firm mounting is important. Try to mount the switch on the underside of a joist. If you can't, fit a batten between the joists so that it sits against the upper surface of the ceiling. Then drill holes in the batten and the ceiling for the cables, remove a knockout from the base of the plastic mounting box and screw it in place. Finally, make up the connections as shown in the illustration.

Left: The new radial circuit cable is run above the ceiling to the 30A double-pole ceiling switch, and is connected as shown. Then a cable is run on to the shower unit, where it is fed in under a cable clamp before being connected to the terminal block. Note the sleeved earth connection to the terminal mounted on the shower unit's metal casing.

1 Feed the circuit cable into the shower unit through the grommet and cable clamp in the base of the case. Prepare the cable cores and sleeve the earth core.

2 Connect the live and neutral cores to the appropriate terminals within the unit and the earth core to its separate terminal. Make sure the terminal screws are tightened fully.

3 Tighten up the cable clamp – here, using long-nose pliers to grasp the nuts. Ideally, a small spanner should be used, but if you don't have one you can use pliers if you are careful. Check that the connections are correct, and replace the shower unit cover.

60 IMMERSION HEATERS

Immersion heaters provide an efficient way of heating water for domestic use when there is no central heating boiler, and they are also often used as an alternative heat source when there is full central heating, since it is more efficient to use the heater in summer than to run the boiler just to generate hot water.

Most immersion heaters — rather like large electric kettle elements — are designed to be fitted into a boss in the top of the hot water cylinder. There are two main types; one has a single element that extends almost to the bottom of the cylinder, while the other has two elements — one long as above, the other about 400mm (16in) long. Both types have to be supplied via a separate radial circuit protected by a

15A fuse and run in 2.5mm² cable (in theory, heaters rated at less than 3kW *could* be plugged into a socket outlet, but would hog that circuit's current-carrying capacity and present a grave risk of overloading it). The circuit cable is run to near the hot cylinder, where a 20A double-pole switch is fitted to control the heater.

In the case of a two-element heater, a 20A dual switch is installed instead; this has a double-pole on-off switch and a changeover switch (marked 'sink' and 'bath') for switching on the appropriate element.

Below: The circuit cable runs to a 20A double-pole switch (or a fused connection unit for heaters rated below 3kW). This is linked to the heater with 1.5mm² heat-resisting flex, connected up as shown.

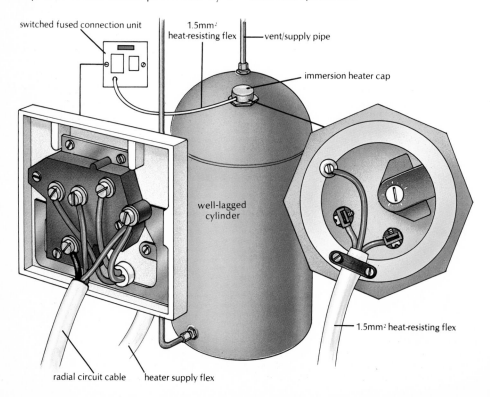

switched fused connection unit   1.5mm²
heat-resisting flex   vent/supply pipe

immersion heater cap

well-lagged
cylinder

1.5mm² heat-resisting flex

radial circuit cable   heater supply flex

SHAVER SUPPLY POINTS 61

You can, of course, plug an electric shaver into a special shaver adaptor that fits into a standard socket outlet, but you won't be able to use your shaver in the bathroom since you are not allowed socket outlets there because of the proximity of water.

It's more convenient to provide special shaver points in rooms where shavers are most commonly used — bathrooms, bedrooms and cloakrooms. These will accept most British, American, European and Australian two-pin shaver plugs.

There are two main types of shaver point: the shaver supply unit and the shaver socket outlet. The former is designed specially for use in bathrooms and washrooms where the proximity of water is so potentially dangerous. Power is provided to the outlet via a transformer, which means there is no direct connection between the shaver and the mains. A self-resetting overload device restricts the power supply to about 20W, so no other appliance could be connected to it without tripping the overload device. The unit usually supplies two voltages, 110V and 240V, selected either via a switch or by using only two of the three available socket holes.

The shaver socket outlet differs from the supply unit in not containing a transformer, so it cannot be installed in a bathroom or washroom. However, it's a much cheaper option for bedrooms. It contains a 1A fuse to protect the shaver, and may also be fitted with an overload device.

Some units are also available combined with a strip light (which is controlled by a pull-cord switch); these come in both shaver supply unit and shaver socket outlet versions. Make sure you get the right type for your chosen location.

You can provide power for a shaver supply unit in two ways: as a spur from a junction box or loop-in rose on a nearby lighting circuit, using 1mm^2 cable or as a spur from a junction box or socket outlet on a ring circuit, using 2.5mm^2 cable (see pages 30 to 33 for details of how to identify and extend lighting and power circuits).

A shaver socket outlet can be supplied as above from a lighting circuit, but if linked to a power circuit the spur must be run via a fused connection unit fitted with a 3A fuse. Then 1mm^2 cable is run on from the fused connection unit to the shaver socket (if you want to install several shaver sockets, simply loop this cable on from one unit to the next).

1 A shaver socket outlet fits a standard single box and can be wired from a lighting circuit. To fit it, connect the cable cores.

2 Tuck the cores neatly into the mounting box, then screw the faceplate into position. Check that it is level.

3 With a shaver supply unit you will need a special deep box if you want the unit to be flush-mounted on the bathroom wall.

62 POWER TO OUTBUILDINGS AND GARDENS

If you want light and power in a detached garage, garden shed or greenhouse, you must install a separate radial circuit to it — you are not allowed to extend your house circuits. There may be a spare fuseway in the consumer unit, which you can use (fit a 30A fuse or MCB), but it's better to install a terminal box and a separate circuit to a new switchfuse unit (see pages 52 to 55), complete with isolating switch and 30A fuse or MCB.

You can run the new sub-circuit in three ways: buried underground, carried overhead or fixed to a boundary or other wall (but *never* to a fence, which could be blown down in high winds). In the first case, you can use ordinary PVC-sheathed cable if the outdoor section is completely contained in steel or rigid plastic conduit; otherwise PVC-covered armoured cable must be used. Mineral-insulated copper-covered (MICC) cable could be used, but the cut ends of the cable have to be fitted with special glands to stop the insulation absorbing moisture, and this needs special tools — an unnecessarily complicated job for the do-it-yourselfer. Where armoured cable is used, a conversion box is fitted at each end of the underground run — one in the house,

Running overhead cables
Below: An overhead cable must be at least 3.5m (11ft) above ground, and 5.2m (17ft) if over a driveway. If the span is over 3.5m, a catenary wire must be fitted to support it. A strainer bolt keeps the wire taut.
Right: Within the outbuilding, the new sub-circuit must be controlled by a switchfuse unit, fed from a terminal box in the house. This then supplies the lighting and power circuits.

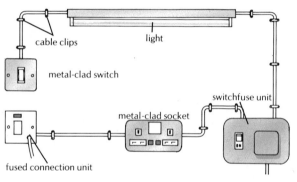

cable clips
light
metal-clad switch
switchfuse unit
metal-clad socket
fused connection unit

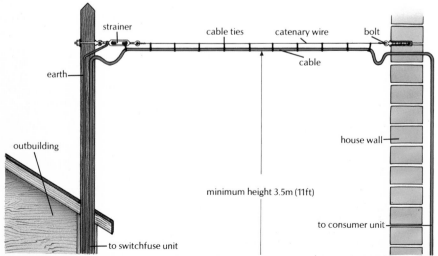

strainer
cable ties
catenary wire
bolt
cable
earth
house wall
outbuilding
minimum height 3.5m (11ft)
to consumer unit
to switchfuse unit

POWER TO OUTBUILDINGS AND GARDENS 63

one in the outbuilding — and the inside sections are run in cheaper ordinary PVC-sheathed cable.

With an overhead cable run, ordinary PVC-sheathed cable can be used unbroken, but certain rules have to be followed. It must be at least 3.5m (11ft) above the ground, 5.2m (17ft) if it crosses a driveway. If the span is over 3.5m (11ft), the cable must be supported on a 'catenary wire', and this catenary must have intermediate supports if more than 5m (16ft 6in) long.

The incoming cable from the terminal box is fed into a switchfuse unit containing one or more fuses or MCBs.

For a small installation, one 30A fuseway will be sufficient; any lighting circuit can be fed via a spur and a 5A fused connection unit. However, it's better to have separate 5A and 30A fuseways for lighting and power circuits. Under new Wiring Regulations, you must fit a 30mA ELCB in the new switchfuse unit to provide additional protection to users of outdoor power tools likely to be run from the new sockets. You can, of course, use any electrical accessories within the outbuilding, but surface-mounted metalclad accessories are the most durable in use.

1 To support an overhead cable run you need a catenary wire. First screw in a straining eye to a tall post fixed in the ground (or to the wall of the house).

2 Fix the catenary cable between post and outbuilding, and link it to the straining eye with an adjustable hook and eye. Take up the slack.

3 You can now run in the supply cable, leaving plenty of slack where it loops up to the catenary. Attach it with cable ties at regular intervals.

4 Within the outbuilding, install a small switchfuse unit with two fuseways: for socket outlets, and for lighting.

5 With metalclad accessories, run in the cable to each box, and add an earth core to link box and faceplate.

6 Sleeve this earth core and link it to the accessory's earth terminal. Connect the other cores and attach the faceplate.

64 POWER TO OUTBUILDINGS AND GARDENS

Unless you want your garden to look like an extension to the national grid, you'll find it much neater to bury long cable runs underground. This will mean digging a trench at least 500mm (20in) deep from where the cable leaves the house to wherever the run terminates. Try to avoid taking the cable across the flowerbeds or vegetable plots, where deep-digging could disturb or damage it; the best place is alongside a path or wall.

If you're laying armoured or MICC cable, put a layer of sand in the bottom of the trench first. Then lay the cable and cover it with a layer of sieved soil before refilling the trench. If you're using conduit lay out the run carefully, then thread the cable through and solvent-weld the couplers to link each length. Cut the last section to length,

add elbows and further short lengths to bring the cable up the house or outbuilding walls, and finish off the run with a further elbow and conduit length to take the cable through the wall.

Where the garden end of the cable is supplying a light or power point, make sure that you install weatherproof fittings; special sealed switches and threaded plug-and-socket outlets are available for outdoor use.

If you're powering pumps and lights for an outdoor pool, you may prefer to use low-voltage equipment and install a transformer in the house or garage. The low-voltage cable, usually at 12V, need not be buried since there is no risk of getting a dangerous shock if you were to accidentally cut the cable. Keep any socket outlets well away from the pool.

Running cables underground
Right: If you're laying cable underground, it must be buried at least 500mm (20in) deep. You can use ordinary PVC-sheathed cable if it's enclosed in conduit (A); armoured cable can be used unprotected, but has to be connected to a conversion box at each end of the run with a special coupler (B).
Below: Low-voltage cable fed from a transformer can be laid on or just below the surface. The transformer itself must be under cover.

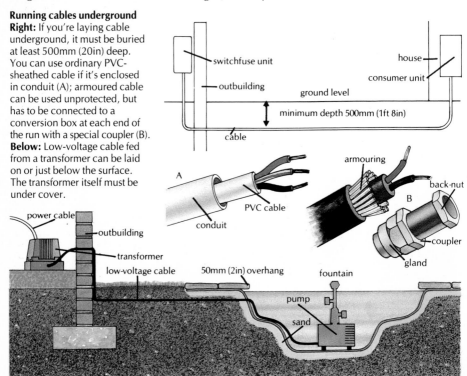

1 You can run ordinary PVC-sheathed cable underground in rigid PVC conduit. Dig a trench and lay out the conduit sections, checking that they align neatly.

2 Thread the cable through the conduit, taking it through elbows where the run changes direction – passing through the house and outbuilding walls, for example.

3 When the cable is threaded through the whole run, work along it, solvent-welding the joins as you go. Brush on the solvent, assemble the joint and twist it to secure a good bond.

4 If you're using armoured cable you won't need conduit. To prepare the cable end for connection to the conversion box, slip on a PVC hood and gland nut.

5 Slip on the gland and coupler, and trim the armouring on the cable so that it ends at the bottom of the thread on the gland. Push up the gland nut and tighten it.

6 Feed the cable end into the conversion box and attach the back-nut to the coupler. Tighten it with a spanner. The coupler allows more room in the box for connections.

7 Attach an earth core to the box, and fit a block of three connectors to join the cores of the armoured and PVC cables.

8 Use PVC-sheathed cable to complete the circuit within the building. Link the box earth to the connector block.

9 Finally connect the other cores within the connector block and fit the lid. A similar box is used in the house.

66 FITTING A DOOR BELL

Most people choose battery-powered door bells because they are simple to install — all you have to do is to fix the bell housing to the wall, the bell push to the front door post, and link the two with a length of twin-core bell wire. But batteries do run down, and won't power an illuminated bell push for long. Instead, use a transformer and run your bell and bell push from the mains.

For this, you need a special bell transformer, usually having three different low-voltage outputs. Which you use depends on the model of bell or chime you have chosen; the manufacturer's instructions will tell you what voltage is needed. From the transformer, run twin-core bell wire to the bell, on to the push and back to the transformer. The transformer is either linked to a spare 5A fuseway in the consumer unit with 1mm² cable, or is linked into a nearby lighting circuit via a three-terminal junction box.

You can add a second bell in series with the first; both will ring when the push is depressed. You can even wire in a change-over switch (an ordinary two-way light switch), as shown, so you can switch off the door bell and switch on a garden bell when you are out of doors.

Connecting up bells
Right: With battery bells the circuit is extremely simple – the bell push simply makes or breaks the circuit. In practice, the battery is within the bell housing, and a length of bell wire links it to the bell push.
Below: Mains bells are run from a 5A fuseway via a transformer. Here, two bells are wired up via a changeover switch, allowing one or the other to be switched off as required. One could be out of doors, or bells could be fitted at both front and back doors.

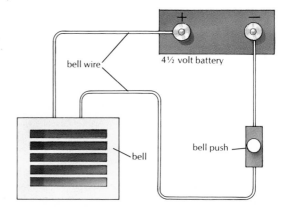

bell wire

4½ volt battery

bell

bell push

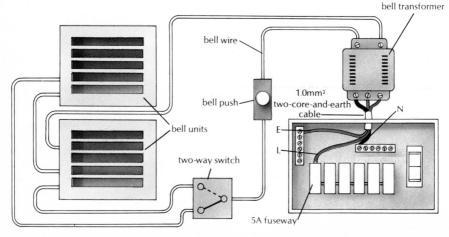

bell transformer

bell wire

bell push

1.0mm²
two-core-and-earth
cable

N

E

L

bell units

two-way switch

5A fuseway

Television and FM radio sets are linked to their aerials with special coaxial cable. This is often run down the outside wall of the house from a chimney stack aerial, entering the house through a hole drilled in a wall or window frame. At that point a small junction box is usually fitted, and a second length of cable with a coaxial plug at each end is used to link the set to the junction box. This looks untidy, and it's much neater to run the coaxial cable inside the house. It can be fed into the roof under a tile near the aerial, and then taken in conduit to the room where the set is installed. The conduit terminates at a TV socket outlet (flush or surface-mounted), and the short TV lead is plugged into it.

When both TV and FM aerials are installed, two downleads will be needed — an added expense. The solution here is to fit a special isolated TV/FM twin socket outlet in the loft, a second one in the living room, and to link the two with just a single down-lead. Each appliance aerial lead is then plugged into its appropriate outlet on the socket faceplate.

Remember that you are not allowed to have a TV socket outlet and a mains socket outlet mounted in a dual box. They must be mounted in separate enclosures, even if for convenience and neatness they are set side by side on the wall.

Central heating controls

The way in which your central heating controls are wired up will depend to a large extent on what controls you have and on the manufacturer's own wiring instructions. However, power is usually provided via a switched fused connection unit containing a 5A fuse and linked as a spur to a ring main. The circuit can also be wired back directly to a spare 5A fuseway in the consumer unit; this means that a failure of the ring main fuse won't affect the heating.

1  2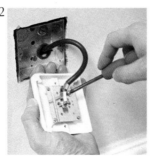

1 Run the coaxial cable from the aerial to the point where you want to install the socket. The cable can be run in conduit buried in the plaster, or can be dropped down wall cavities.
2 Strip back the sheathing and screening wires; clamp these in the special clip and then connect the central core to the aerial terminal. Finally, attach the faceplate to the box.

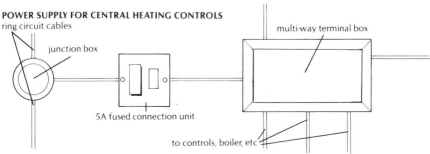

POWER SUPPLY FOR CENTRAL HEATING CONTROLS
ring circuit cables
junction box
multi-way terminal box
5A fused connection unit
to controls, boiler, etc

68 REWIRING YOUR HOUSE

Some of the tell-tale signs that your house needs rewiring have already been mentioned, and if you have been contemplating any electrical work you may already be aware of the short-comings of your wiring system. Here's a checklist of the danger signs:

Under the stairs . . .
● a multitude of separate old-fashioned switchfuse units
● double-pole fusing — a fusebox with fuses in both the live and neutral poles.

Around the house . . .
● round tumbler switches mounted on wooden pattresses
● round-pin socket outlets on pattresses or skirting boards
● old-fashioned ceiling roses, and pendant lights hanging from twisted twin flex
● adaptors being used to make up for a shortage of sockets
● overheating at sockets.

Behind the scenes . . .
● single-core cables run in light-gauge metal conduit, often rusty
● rubber or lead alloy sheathed cables
● crumbling core insulation behind switches and sockets
● missing earth continuity conductors on circuits.

You may find that parts of your system *have* been rewired — for example, a new cooker circuit or a ring main may have been installed. If these extensions were done properly, with PVC-sheathed cable run back to new switch-fuse units, they can probably be left as part of the new system. However, it's best to have them checked by your local electricity board or a qualified electrician first. Don't be deceived by the presence of new cable (which may just be a spur added to an existing — and dangerous — circuit) or new accessories (a popular and purely cosmetic dodge often perpetrated by house sellers); such recent additions

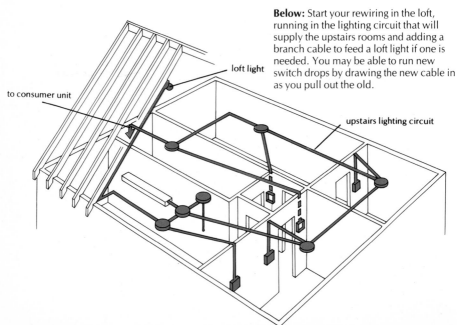

Below: Start your rewiring in the loft, running in the lighting circuit that will supply the upstairs rooms and adding a branch cable to feed a loft light if one is needed. You may be able to run new switch drops by drawing the new cable in as you pull out the old.

to consumer unit

loft light

upstairs lighting circuit

may be reusable, but only after proper rewiring has been carried out.

The actual task of rewiring a house is not especially difficult; it's really the sum of all the techniques and projects already described. However, it's a time-consuming and disruptive job best carried out in an empty house, or in planned stages in an occupied one. It needs some careful estimating for materials, too.

The first stage is to work out exactly where you are going to need lighting points, socket outlets and special provision for accessories such as cookers and immersion heaters, central heating electrics, showers and circuits to outbuildings. Start with a sketch plan of your house, and mark in everything you can think of. Then start to break the system up into circuits; remember, a lighting circuit should feed a maximum of twelve (and an optimum of eight) lighting points, and a ring main a maximum floor area of 100sq m

(1076sq ft). In a typical home, your circuit requirements on a well-planned system could be:

Requirement	Rating
Upstairs lighting circuit	5A
Downstairs lighting circuit	5A
Radial circuit to immersion heater	15A
Radial circuit to instantaneous shower	30A
Radial circuit to outbuilding	30A
Ring circuit to first floor	30A
Ring circuit to ground floor (except kitchen)	30A
Ring circuit to kitchen only	30A
Radial circuit to cooker	45A

That's nine circuits already, not including 'luxury' circuits for freezers, mains bells or central heating electrics. This total will decree what size of consumer unit you need; remember to allow for

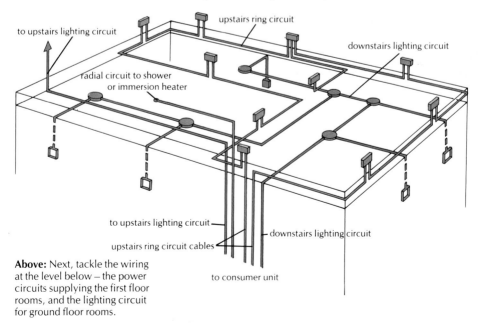

to upstairs lighting circuit

upstairs ring circuit

downstairs lighting circuit

radial circuit to shower or immersion heater

to upstairs lighting circuit

downstairs lighting circuit

upstairs ring circuit cables

to consumer unit

Above: Next, tackle the wiring at the level below – the power circuits supplying the first floor rooms, and the lighting circuit for ground floor rooms.

70 **REWIRING YOUR HOUSE**

further room in it if you want an integral ELCB instead of a mainswitch as an isolator unit. Choose a unit fitted with MCBs, or possibly cartridge fuses, not rewirable fuses.

The next job is to plan the room-by-room requirements in more detail so you can total the number of electrical accessories of each type you will need. Do not forget to include mounting boxes as required.

Thirdly, you need a rough estimate of the amount of cable to buy. There's little point in measuring cable runs accurately except on heavy-duty circuits such as those to cookers and showers, where the cable is very expensive; simply buy 50m drums of $2.5mm^2$ and $1mm^2$ two-core and earth cable for power and lighting circuits respectively, plus shorter lengths of other cable types required ($4mm^2$, $1mm^2$ three-core and earth for two-way switching, etc.). You can always buy another drum, or cable by the metre, later, should you run out. Add a quantity of $1mm^2$ round three-core flex for pendant lights, heat-resisting flex for immersion heaters, and other equipment and sundries such as PVC channelling or conduit, BESA boxes and cable clips. Then take your full shopping list to at least two wholesalers and ask for a quotation. Make sure that all accessories are made to the appropriate British Standards (BS 1363 for socket outlets, 5733 for connection units, 3676 for switches, 4177 for cooker control units, 3052 for shaver supply units, 67 for ceiling

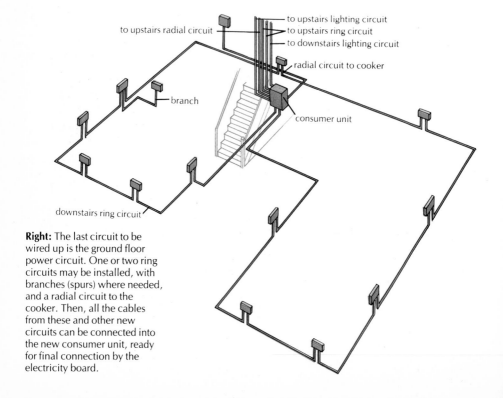

to upstairs radial circuit
to upstairs lighting circuit
to upstairs ring circuit
to downstairs lighting circuit
radial circuit to cooker
branch
consumer unit
downstairs ring circuit

Right: The last circuit to be wired up is the ground floor power circuit. One or two ring circuits may be installed, with branches (spurs) where needed, and a radial circuit to the cooker. Then, all the cables from these and other new circuits can be connected into the new consumer unit, ready for final connection by the electricity board.

REWIRING YOUR HOUSE 71

roses, 6004 for cables and 6500 for flexes).

The actual job is best split up into stages to minimise disruption. Start in the loft, wiring up the bedroom lighting circuits and adding extras such as loft lights and TV aerial sockets. Then move to the next level down — the void between first and ground floors — and tackle the ground floor lighting circuit and the first floor power circuits. Finally, tackle the ground floor power circuits (unless you have solid ground floors, in which case these will also be run at ceiling level, with vertical chases in the walls carrying cable down to the socket outlets).

Throughout, chop out and fix mounting boxes first, so you can measure and cut cable runs precisely. On an exten-sive rewire, you're likely to be choosing new positions for most accessories anyway, so you can leave existing items working as you run in the new circuits. Then run all your circuit cables neatly back to the new consumer unit position, which doesn't *have* to be next to the meter. It may be more convenient on a hall or kitchen wall, and is simply linked via a sub-main cable to a new main isolator switch next to the meter.

At this point, you will have to notify your local electricity board to come and disconnect the old switchgear, connect up your new consumer unit and test the installation. This is the moment of truth, where good workmanship and careful planning will have resulted in a sound and safe job.

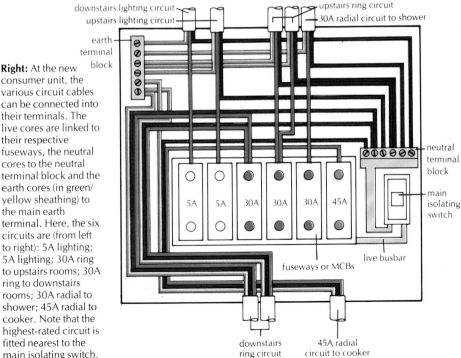

Right: At the new consumer unit, the various circuit cables can be connected into their terminals. The live cores are linked to their respective fuseways, the neutral cores to the neutral terminal block and the earth cores (in green/yellow sheathing) to the main earth terminal. Here, the six circuits are (from left to right): 5A lighting; 5A lighting; 30A ring to upstairs rooms; 30A ring to downstairs rooms; 30A radial to shower; 45A radial to cooker. Note that the highest-rated circuit is fitted nearest to the main isolating switch.

downstairs lighting circuit
upstairs lighting circuit
upstairs ring circuit
30A radial circuit to shower
earth terminal block
neutral terminal block
main isolating switch
live busbar
fuseways or MCBs
5A 5A 30A 30A 30A 45A
downstairs ring circuit
45A radial circuit to cooker

72 PLUGS AND FUSES

Because many homes still have round-pin sockets, electrical appliance manufacturers usually sell their equipment without a plug (and most shops expect you to pay extra for one when you buy an appliance, even though you can't use it without one). So you will often need to fit a plug to any new appliance you buy, and you may also have to replace an existing plug that has been damaged. Because the plug is such a vital link between the appliance and the mains, it is vital that you do this properly. There are several rules to remember, whatever type of plug you are fitting:

1 colour coding of flex cores In appliances (and flex) made since about 1970, the live core is sheathed in brown PVC, the neutral core in blue PVC and the earth in green/yellow striped PVC. Earlier equipment and older flex had the same colour codes as mains cable — red for live, black for neutral and green for earth; make sure you know which is which. On continental appliances, check with the manufacturer or his agent if the colour coding is different from this.

2 terminals With the plug open and the inside facing you, the top terminal linked to the larger pin is the earth

1 To connect new flex to a plug, first split and cut away about 38mm (1½in) of the flex sheathing. Take care not to cut through the insulation on the inner cores.

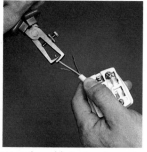

2 Lay the cable in the plug and cut each core long enough to reach its appropriate terminal when the flex sheathing is positioned over the cable grip. Strip each core.

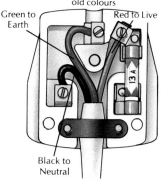

old colours

Green to Earth

Red to Live

13 A

Black to Neutral

3 Clamp the flex sheathing in the cable clamp, and then link each core to the appropriate terminal as shown. Check that the terminal screws are tight.

4 Clip in a cartridge fuse of the correct rating for the appliance. Then check all connections again before fitting the top and tightening the screw.

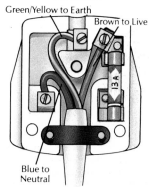

new colours

Green/Yellow to Earth

Brown to Live

13 A

Blue to Neutral

Old flex cores are a different colour to new flex cores, as shown above.

(usually marked E or ⏚). The bottom right terminal (marked L) is live and the bottom left (marked N) is neutral. Remember that the BRown core goes to the Bottom Right terminal, the BLue core to the Bottom Left terminal.

3 insulation There should be no bare strands visible inside the plug; the core insulation should reach right up to the terminal. The outer sheathing should be clamped securely in the cord grip.

4 two-core flex This is used only on double-insulated appliances (marked ▣) and certain table lamp fittings with plastic lampholders. In this case, there is no earth core; the other cores are connected as above.

5 fuses In fused plugs, fit a brown 13A fuse for appliances rated at over 700W (and for colour televisions), a red 3A fuse otherwise. You will find the appliance rating stamped on or attached to its body. Test whether fuses have blown with a continuity tester, or by holding them across the open end of a

metal-cased torch with one end of the fuse on the casing and the other on the end of the battery. If the fuse is intact, the torch will work when switched on.

Mending a circuit fuse

When a circuit fuse blows, there is no point in mending it until you have tracked down and cured the fault that caused it to blow in the first place (see page 75). It may be obvious which circuit fuse has blown, but if not, switch off the main isolator switch and remove each fuse from its holder in turn. If you have rewirable fuses, you will be able to see at a glance which one is at fault, but with cartridge fuses you will probably have to test each fuse with a continuity tester (or by using the torch trick explained previously). Mend the fuse, then replace the other fuses and restore the power.

If you have MCBs, all you have to do to restore the power is to press the reset button or move the switch to ON.

1 To replace blown fuse wire in a rewirable fuse, loosen the terminal screws and remove the remains of the old wire. Then thread in new wire of the correct rating.
2 Cut off the excess, and connect each end to the terminals, winding it round once. Don't pull the wire taut. Tighten each terminal screw and then replace the fuse.
3 With cartridge fuses, simply open the fuse carrier, take out the old fuse and insert a new one of the same rating.
4 You can easily test whether cartridge fuses are intact by holding them across the open end of a metal-cased torch, one end touching the battery and the other resting on the torch body. With the torch switched on, the bulb will light if the fuse is intact.

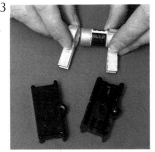

Fit new flex to an appliance as soon as it shows any signs of damage or wear. You are more likely to get a shock or cause a short circuit from faulty flex than from almost any other fault in the electrical system, since live cores can so easily be exposed.

It's important to choose flex of the correct type and current rating for the job it has to do. Choose two-core flex for double-insulated appliances (marked with a double-square symbol ▣), and for pendant and table lamps with non-metallic fittings; go for three-core flex in all other cases. Use ordinary PVC-sheathed flex for most jobs, braided flex for electric heaters and fires, unkinkable flex for portable kitchen appliances such as irons and kettles, and heat-resisting flex in light pendants with bulbs over 100W and also for wiring up immersion heaters. Choose the size of flex according to the appliance rating:

Flex size (mm^2)	0.5	0.75	1.00	1.5
Appliance rating (W)	700	1400	2400	3000

NEVER extend flex except with a proper one-piece flex connector. If you are using an extension plug-and-socket, connect the male part (with the pins) to the appliance, the female part to the mains — NEVER the other way round.

1 To replace a flex, unplug the appliance, and open the casing to reveal the terminals. Disconnect the flex, saving any protective sleeving.

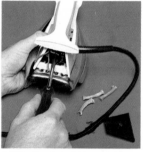

2 If the flex is held by a cable clamp, undo this and then draw the old flex out through the grommet attached to the appliance casing.

3 Feed the new flex in through the grommet. It's easier to do this before you strip and prepare the cable ends.

4 Finally, connect the new flex cores to the terminals within the appliance, reusing any protective sleeving.

Extending flex safely
Below: If you're extending flex with a two-part plug and socket, ALWAYS link the plug part to the appliance flex and the socket part to the mains. Otherwise, the pins will be live if the parts are separated.

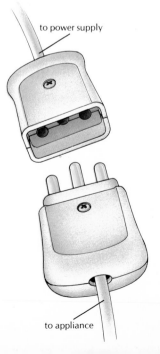

to power supply

to appliance

When something electrical fails to work around the house, the fault is usually fairly simple to locate and cure. With a knowledge of your electrical system and how it works, all you have to do is check things in a logical order until you pin-point the problem.

Turn off the power at the fusebox or consumer unit before investigating any faults on fixed appliances or circuit wiring; unplug any portable appliance before starting work on it.

A Pendant light doesn't work
1 turn off light switch, replace bulb.
2 turn off main switch, check lighting circuit fuse/MCB, replace/reset if necessary, restore power; if fuse blows again or light won't work, go to 3.
3 with power off, open rose and lampholder at offending light, look for loose connections or broken cores, and strip and remake connections as necessary; check that flex cores are hooked over anchorages, replace covers and restore power; if fuse blows again or light won't work, go to 4.
4 with power off, disconnect pendant flex and use continuity tester to check each core in turn; replace the flex if any core fails, replace rose/lampholder covers and restore power; if fuse blows again, go to FAULT C.

B Electrical appliance doesn't work
1 plug in appliance at another socket; if it works, suspect fault at original socket (FAULT C); if it doesn't, go to 2.
2 fit new plug fuse of correct rating.
3 check flex connections at plug terminals and remake if necessary.
4 with appliance unplugged, open casing and check connections at terminal block; remake if necessary.
5 check appliance flex continuity as in A4, and replace flex if necessary with new flex of correct current rating.
6 if checks 2 to 5 fail, suspect circuit fault (FAULT C) or appliance failure.

C Whole circuit is dead
1 switch off all lights/disconnect all appliances linked to affected circuit.
2 replace circuit fuse (using fuse/wire of the correct rating) or reset MCB.
3 switch on lights/plug in appliances one by one; note which one causes the fuse to blow/MCB to trip, then:
4 isolate the offending light/appliance (and check as in FAULTS A and B), then replace fuse/reset MCB again.
5 if circuit is still dead, check switches and socket outlets on circuit for physical damage or faulty connections at terminals, and replace or reconnect.
6 replace damaged circuit cable if pierced by drill or nail.
7 if the circuit fault persists, call in a qualified electrician.

D Whole house system is dead
1 check with neighbours to see if there is a local power cut.
2 if an ELCB is fitted to the system, check whether it has tripped off, and reset it if necessary; if it cannot be reset, there is a fault somewhere on the system. Run through FAULTS A, B and C, or call an electrician.
3 call electricity board to check main service fuse (you must not tamper with this yourself).

E Electric shock received
1 if the shock was minor, isolate the offending appliance for checking or replace the accessory concerned.
2 if someone receives a major shock, try to turn off the power, grab their clothes (NOT bare flesh, or you will get a shock if the power is still on) and drag them away from the power source; administer artificial ventilation or external chest compression — see page 76 for more details — and call a doctor or ambulance.
3 have the fault tracked down immediately and put right by an electrician; a repeat could kill.

76 FIRST AID FOR ELECTRIC SHOCK

If you or a member of your family receive a slight shock from an appliance, STOP USING IT. Unplug it immediately; check the connections at the plug (page 72) and terminal block (page 74), remaking them if necessary, and check and replace the flex, too, if this is suspect (page 74). As an extra safety precaution, have it checked over by a qualified electrician. If the shock is received from a switch, socket outlet or other accessory, call a qualified electrician to check the circuit and look for any other faults that may be present.

If someone receives a severe electric shock, they may involuntarily grip the source of the current. IMMEDIATELY turn off the current if a switch is nearby, or drag them away by their clothing. DON'T TOUCH their flesh or you will receive a shock, too.

If conscious, but visibly shocked (deathly pallor, sweating, rapid and shallow breathing), lay the casualty flat on the back on a blanket with legs raised slightly on a pillow, unless you suspect leg fractures. Turn the head to one side to keep the airway clear, and cover the patient with a blanket. DON'T apply a hot water bottle. DON'T move the patient unnecessarily, nor give them anything by mouth (if thirsty, moisten the lips with water), nor allow them to smoke.

Cool any burns by flooding them gently with cold water. Then cover with a sterile dressing (a clean pad of non-fluffy material will do), securing it with a bandage. DON'T apply ointments, lotions or fats to the burn, and DON'T break any blisters or remove loose skin. CALL AN AMBULANCE.

If unconscious, lay the casualty in the recovery position shown below, placing the limbs so that they support the body in this position. Keep the airway clear by tilting the head back and bringing the jaw forwards. Cover with a blanket and CALL AN AMBULANCE IMMEDIATELY. Watch for signs of cessation of breathing and heartbeat, giving artificial ventilation and external chest compression respectively, as necessary (see opposite).

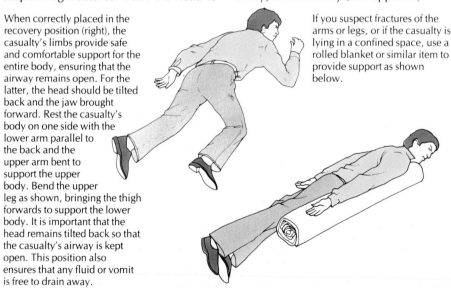

When correctly placed in the recovery position (right), the casualty's limbs provide safe and comfortable support for the entire body, ensuring that the airway remains open. For the latter, the head should be tilted back and the jaw brought forward. Rest the casualty's body on one side with the lower arm parallel to the back and the upper arm bent to support the upper body. Bend the upper leg as shown, bringing the thigh forwards to support the lower body. It is important that the head remains tilted back so that the casualty's airway is kept open. This position also ensures that any fluid or vomit is free to drain away.

If you suspect fractures of the arms or legs, or if the casualty is lying in a confined space, use a rolled blanket or similar item to provide support as shown below.

FIRST AID FOR ELECTRIC SHOCK 77

1 Before commencing mouth-to-mouth ventilation, turn the casualty's head to one side and remove any obstructions from the mouth with your index finger. Don't delay doing this as you must begin ventilation immediately. Then open the casualty's airway by tilting the head backwards and pushing the chin upwards.

2 Making sure the casualty's head is tilted back to keep the airway open, take a deep breath and pinch the casualty's nostrils closed with your finger and thumb.

3 Seal your lips around the casualty's mouth and blow into it. Watch for the casualty's chest to rise, remove your mouth, exhale any excess air, and watch the chest fall. Repeat

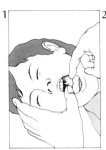

the process, making the first four breaths as rapid as possible. Check for a pulse at the casualty's neck by placing the tips of your fingers in the hollow of the neck between the voice box and adjacent muscle – just to one side of the throat. Continue at normal breathing rate until the casualty is

breathing normally. Then place the casualty in the recovery position (opposite). If mouth-to-mouth ventilation is unsuccessful on its own (casualty has no pulse, is deathly pale and has developed a blueness around the mouth and ear lobes), it should be used with external chest compression (below) at once.

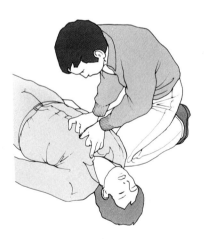
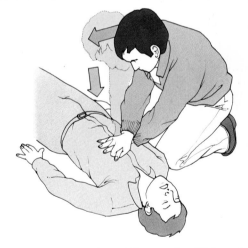

First, locate the centre of the lower half of the breast bone by placing your thumbs midway between the sternal notch at the top of the chest and the intersection of the rib margins at the bottom (above, left). Place your hands as shown at this mid point (above, right), locking the fingers together and keeping them clear of the patient's ribs.

With your arms straight, position yourself above the casualty until they are vertical and then press down 40-50mm (1½-2in). Release the

pressure and repeat the procedure 15 times at a rate of 80 per minute. Give two full breaths of mouth-to-mouth ventilation (above) and then continue the compression procedure for one minute.

Check for heartbeat (as described above) and then continue the compression/ventilation cycle described, checking the heartbeat every three minutes. When the heartbeat returns, continue mouth-to-mouth ventilation until the casualty is breathing normally.

amp short for ampere (A), used to measure the flow of electricity through a circuit or appliance.

cable conductors (or 'cores') covered with a protective semi-rigid insulating sheath, used to wire up the individual circuits on a wiring system.

circuit any complete path for an electric current, allowing it to pass along a 'live' conductor to where it's needed, and then to return to its source along a 'neutral' conductor.

conductor the metallic current-carrying 'cores' within cable or flex.

consumer unit unit governing the supply of electricity to all circuits, and containing a main on-off switch and fuses or circuit breakers protecting the circuits emanating from it.

earthing the provision of a continuous conductor on circuits to protect the user from certain electrical faults. Earth conductors are sheathed in green/yellow striped PVC; earth terminals are marked E or ⏚ .

earth-leakage circuit breaker (ELCB) device fitted to circuits to detect current leakage that could start a fire or cause an electric shock.

flex short for 'flexible cord' — cores wrapped in a flexible outer sheath, used to link appliances and lights to the house's fixed wiring circuits.

fuse protective device designed to cut off the flow of current in a circuit in the event of a fault.

gang used to describe the number of units — switches or socket outlets — contained in one electrical accessory. A two-gang switch has two switches mounted on one faceplate.

insulation the PVC sheathing on and around the cores of cable and flex.

lamp correct 'trade' term for a light bulb or tube.

live used to describe the cable or flex core taking current to where it is needed, or any terminal or part to which the core is connected. Live

cable cores are colour-coded in red, live flex cores in brown. Live terminals are marked L.

miniature circuit breaker (MCB) a device used instead of fuses to isolate a circuit.

neutral used to describe cores carrying current back to its source, or any terminal to which the core is connected. Neutral cable cores are colour-coded black, neutral flex cores are blue. Neutral terminals are marked N.

one-way used to describe a switch with on-off control of the circuit to which it is fitted from that one point only. Two-way switches, used in pairs, offer control of the same circuit from two positions.

radial circuit a power or lighting circuit originating at the consumer unit and feeding one or more electrical accessories, terminating at the remotest one.

ring circuit a power circuit wired as a continuous loop, both ends of which are connected to the same terminals of the consumer unit.

single pole used to describe a switch that 'breaks' only the live side of the circuit it controls. A double-pole switch breaks both the live and neutral sides of the circuit.

spur a sub-circuit run as a branch line from an existing circuit to extend the number of accessories supplied.

unit a measure of the amount of electricity consumed by an appliance or system, the product of the power consumed (in watts) and the time during which it was supplied. One unit is 1 kilowatt-hour (kWh) — used by, for example, a 100W light bulb burning for 10 hours.

volt unit of electrical pressure (potential) difference. In most British homes, the mains voltage is 240V.

watt unit of power consumed by an appliance or circuit, the product of the mains voltage and the current drawn (in amps). 1000W = 1 kilowatt (kW).

Acknowledgements
Tools, Equipment & Facilities

The publishers are grateful to the many organisations and individuals who supplied materials, tools and other equipment, or provided locations and/or facilities for photography. Thanks are due especially to the following:

Ashley Accessories Ltd, Ulverston, Cumbria (Mr Alan Brook): Electrical fittings.

B & R Electrical Products Ltd, Harlow, Essex (Mr Peter Barnett): ELCB socket and PowerBreaker plug.

Blomberg (UK) Ltd, Birkenhead, Merseyside (Elizabeth King, Link Communications): Electric cooker.

City Electrical Factors Ltd, Stoke Newington, London (Mr Brian Vince): Electrical fittings, electrician's tools.

Crabtree Electrical Industries Ltd, Walsall, West Midlands (Mr Alan Preston): Electrical fittings.

Econa Appliances Ltd, Solihull, West Midlands (Elizabeth King, Link Communications): Parkamatic Silver waste disposal unit.

General Woodwork Supplies Ltd, Stoke Newington, London (Mr Geoff Bentley): General purpose tools.

Home Automation Ltd, Hoddesdon, Herts (Mrs Weston): Dimmer switches.

Marlin Lighting Ltd (including CONELIGHT), Feltham, Middlesex (Mr C R Fielding): Light fittings.

Marshall-Tufflex Ltd, Hastings, East Sussex (Mr D E Smart/Mr J C Boles): Electrical conduit and trunking.

Philips Electrical Industries Ltd (Lighting Division), Croydon, Surrey (Mrs Pam Gillett): Lamps.

Rock Electrical Accessories Ltd, Brentford, Middlesex (Mr Brian Godwin): Electrical fittings.

Thorn-Emi Lighting Ltd, Tottenham, London (Mr Cyril Phillips): Lamps.

Tools and equipment co-ordinator Mike Trier.

Illustrations The photographs and artwork in this book were specially commissioned from the following, to whom the publishers extend their thanks:
Photography **Jon Bouchier** 11, 24-74; **Simon de Courcy Wheeler** 16-21.
Artwork **Chris Forsey** 76-77; **Brian Watson/Linden Artists** 6-74.

#18961357

Software

Electric C

Software fo

It provides

great aid to

Software is

IBM PC and

To receive

    complet

    ESM, Du

Telephone

Please sen

  ☐ BBC

I enclose m

Please cha

  ☐☐☐

Name_____

Address __

_____

_____

This book is to be returned on or before
the last date stamped below.

17 DEC 1990   0 2 FEB 1995

30 MAY 1991 1 8 MAR 1996

20.9.91   1 5 DEC 1999

2 6 NOV 1991   1 8 FEB 2000

2 7 JAN 1992

DUE

1-9 JUN 2008

LIBREX

Software Order Form

Computer Circuit Theory

Software forms an essential part of this Computer Illustrated Text. It provides worked examples, calculating power where necessary and is a of value to understanding and learning.

Software is available for BBC Master, ADFS 40 track 5¼ disc and for and compatibles on 5¼ disc at £34. flat disc.

Texts use Psion software.

Complete this form and return with your remittance to:

TCM, Duke Street, Ambach, Cambs PE13 2AE. Telephone 0945 9341

Remaining orders are associated from customers paying by credit card.

Please send me software for Electric Circuit Theory by B E Hindes on:

BBC Master ADFS 40 track ☐   IBM PC disc ☐

I enclose my cheque for £ _____ payable to £34

Please charge to my Access/Visa card number

☐☐☐☐ ☐☐☐☐ ☐☐☐☐ ☐☐☐☐   Expiry date _____

Name _____

Signature _____

Address _____

Date _____